"For too long, academia, as well as international policy communities, have spoken in vague terms regarding nuclear modernization. This book provides a much needed, and long overdue, conceptual and comparative study of nuclear modernization in the United States, Russia and China. At a time when so many nuclear weapons states are modernizing their arsenals, and arms control seems to be in decline, this is a must-read for nuclear scholars and practitioners."

—**Nicola Leveringhaus**,
King's College London, UK

"Warren and Baxter's new collection of essays is good news for scholars, students, and the policy community. The clear-eyed and concisely written chapters provide a rock solid survey of nuclear modernization underway in the US, Russia, and China. The diverse, mostly younger group of scholars and practitioners also examines the implications nuclear modernization may have for nonproliferation, for international security more generally, and perhaps even for international relations. The ladder may seem unlikely, but in a world where efforts to limit and manage nuclear arsenals are being tossed aside, and where the three largest nuclear powers eye the actions of others with suspicion, today's nuclear modernizations may yet prove to be a pathway to a very different future. Some analyst would welcome that outcome, even as others offer warnings, but all of them will find *Nuclear Modernization in the 21st Century* a useful guide to an uncertain future."

—**James J. Walsh**,
MIT, Massachusetts, USA

"As a new nuclear arms race looms, this important and timely volume assembles leading experts to provide cutting-edge analyses of the plans and programs of the United States, Russia and China to spend billions of dollars on the modernization of their nuclear arsenals. This will be the go-to volume for understanding the new weapons technologies of the major nuclear-weapons states, as well as the likely impact of these ever more sophisticated and dangerous arsenals on global security, stability, and the prospects for proliferation. It will be essential reading for policymakers and academics alike."

—**Nina Tannenwald**,
Brown University, USA

Nuclear Modernization in the 21st Century

This collection examines the extent to which nuclear weapons modernization has become a significant point of concern and consideration in international security. Recent statements and substantial investments by nuclear weapon possessor states in the upkeep and modernization of their nuclear postures – particularly the United States, Russia and China – illustrate a return of primacy and the salience of nuclear forces in international politics. The upgrading of systems, the introduction of new capabilities, the intermingling of new technologies, and the advancement of new strategic models, are all indicative of their elevation in importance and reliance.

With contributions from leading thinkers in the nuclear weapons domain, this book elucidates the global strategic and policy implications such modernization efforts by the above-mentioned states will have on international security. In unpacking and conceptualizing this developing source of potential (in)security and tension, the collection not only provides a technical context, but also frames the likely effects modernization could have on the relations between these nuclear weapon powers and the larger impact upon efforts to curb nuclear weapons – both in terms of horizontal and vertical proliferation. The chapters have been arranged so as to inform a variety of stakeholders, from academics to policy-makers, by connecting analytical and normative insights, and thereby, advancing debates pertaining to where nuclear modernization sits as a point of global security consternation in the 21st century.

Aiden Warren is Associate Professor of International Relations at RMIT University, Melbourne, Australia. He is the 2018–19 Fulbright Scholar in Australia-United States Alliance Studies, sponsored by the Australian Government's Department of Foreign Affairs & Trade (DFAT).

Philip M. Baxter is a Research Fellow with the Center for Policy Research at the University of Albany, founder and director of a data analytics and consultancy firm, and a PhD candidate in International Affairs, Science, and Technology at the Georgia Institute of Technology.

Modern Security Studies
Series editors: Sean S. Costigan and Kenneth W. Estes

This series fills a known gap in modern security studies literature by pursuing a curated, forward-looking editorial approach on looming and evergreen security challenges. Short and long form works will be considered with an eye towards developing content that is widely suitable for instruction and research alike. Works adhere around the series' four main categories: Controversies, Cases, Trends and Primers. We invite proposals that pay particular attention to controversies in international security, notably those that have resulted in newly exposed and poorly defined risks to non-state legitimacy, international or state capacities to act, and shifts in global governance. Case studies should examine recent historical events and security-related actions that have altered present day understanding or political calculations. Trends will need to detail future yet tangible concerns in a 5–10 year timeframe. Authors are also invited to submit proposals to our primers category for short form works on key topics that are referenced and taught throughout security studies.

Verifying Nuclear Disarmament
Thomas Shea

NATO's Democratic Retrenchment
Hegemony after the Return of History
Henrik B. L. Larsen

The Politics and Technology of Cyberspace
Danny Steed

Post-Cold War Anglo-American Military Intervention
A Study of the Dynamics of Legality and Legitimacy
James F.D. Fiddes

Nuclear Modernization in the 21st Century
Edited by Aiden Warren and Philip M. Baxter

For more information about this series, please visit: www.routledge.com/politics/series/ASHSER1437

Nuclear Modernization in the 21st Century

Edited by Aiden Warren and
Philip M. Baxter

First published 2020 by Routledge

2 Park Square, Milton Park, Abingdon, Oxon OX14 4RN
605 Third Avenue, New York, NY 10017

Routledge is an imprint of the Taylor & Francis Group, an informa business

First issued in paperback 2022

Publisher's Note

The publisher has gone to great lengths to ensure the quality of this reprint but points out that some imperfections in the original copies may be apparent.

British Library Cataloguing-in-Publication Data
A catalogue record for this book is available from the British Library

Library of Congress Cataloging-in-Publication Data
A catalog record for this book has been requested

ISBN: 978-1-138-35055-7 (hbk)
ISBN: 978-1-03-233695-4 (pbk)
DOI: 10.4324/9780429435829

Typeset in Times New Roman
by Apex CoVantage, LLC

Contents

List of contributors ix
Foreword xiv
Acknowledgments xvi
List of acronyms xvii

1 **Introduction** 1
AIDEN WARREN AND PHILIP M. BAXTER

2 **U.S. modernization efforts and the 2018 *Nuclear Posture Review*** 12
HANS M. KRISTENSEN

3 **Russia's nuclear modernization** 25
PHILIP M. BAXTER

4 **U.S.-Russian bilateral disarmament** 40
CAMERON TRAINER

5 **Chinese nuclear strategy** 61
SUSAN TURNER HAYNES

6 **U.S. nuclear weapons modernization and the impact on the nuclear nonproliferation regime** 86
AIDEN WARREN

7 **NATO nuclear modernization** 106
STEVEN PIFER

8 **Through a crystal ball, dimly: nuclear modernization's anticipated effects on International Relations theory** 124
BALAZS MARTONFFY AND ELENI EKMEKTSIOGLOU

9 **Modernization a determent to international security** 145
KINGSTON REIF AND ALICIA SANDERS-ZAKRE

10 **Modernization as a promoter of international security: the special role of U.S. nuclear weapons** 173
MATTHEW KROENIG AND CHRISTIAN TROTTI

11 **Afterword** 191
PHILIP M. BAXTER AND AIDEN WARREN

Index 198

Contributors

Editors

Aiden Warren is Associate Professor of International Relations at RMIT University, Melbourne, Australia. He is the 2018–19 Fulbright Scholar in Australia-United States Alliance Studies, sponsored by the Australian Government's Department of Foreign Affairs & Trade (DFAT). Dr. Warren's teaching and research interests are in the areas of International Security, U.S. national security and foreign policy, U.S. Politics (ideas, institutions, contemporary and historical), International Relations (especially great power politics), and issues associated with Weapons of Mass Destruction (WMD) proliferation, nonproliferation and arms control. He has spent extensive time in Washington D.C. completing fellowships at the James Martin Center of Nonproliferation, the Arms Control Association (ACA), and Institute for International Science and Technology Policy (IISTP) at George Washington University. Dr. Warren is the sole author of *The Obama Administration's Nuclear Weapon Strategy: The Promises of Prague* and *Prevention, Pre-emption and the Nuclear Option: From Bush to Obama*; and co-author of *Governing the Use-Of-Force in International Relations: The post 9/11 U.S. Challenge on International Law*, *Presidential Doctrines*, and *Weapons of Mass Destruction: The Search for Global Security*. He is Editor of *Rethinking Humanitarian Intervention in the 21st Century*, and is also the Series Editor of the Weapons of Mass Destruction (WMD) book series with Rowman and Littlefield, New York.

Philip M. Baxter is a Research Fellow with the Center for Policy Research at the University of Albany, founder and director of a data analytics and consultancy firm, and a PhD candidate in International Affairs, Science, and Technology at the Georgia Institute of Technology. His dissertation examines the structure of epistemic communities in nuclear weapon proliferating states. He also serves as a Lecturer of International Affairs at the Elliott School of International Affairs at George Washington University, where he teaches courses on international security and emerging technology. He has over a decade of experience in the international security domain, having worked previously at the James Martin Center for Nonproliferation Studies, National Nuclear

Security Administration, and the National Defense University. His research has appeared in *International Areas Studies Review*, *Journal of Cybersecurity, Science and Diplomacy*, Federation of American Scientists' *Public Interest Reports*, *Arms Control Wonk*, and *Real Clear Defense*.

Contributors

Eleni Ekmektsioglou is a PhD candidate and an adjunct instructor at American University's School of International Service, as well as a pre-doctoral fellow at the Institute for Security and Conflict Studies within the Elliott School of International Affairs at George Washington University. Her dissertation seeks to explain variation in state reactions to new military technologies. Her broader research interests include technological innovation in strategic weapon systems and their impact on crisis stability, international security in East Asia and China's nuclear and naval strategy. Before starting her doctoral studies at American University, Eleni was a research fellow at Pacific Forum CSIS (Center for Strategic & International Studies). She has also worked as a project coordinator at the European Union Institute for Security Studies (EUISS) and at the European Institute for Asia Studies (EIAS). She holds a Master's degree from the King's College London War Studies department and a Bachelors degree from Panteion University of Athens while she has spent one year in Sciences Po Lille as an exchange student. She has written opinion pieces for online magazines such as *The Diplomat* or the *National Interest* and more extensive analyses for journals such as the *Strategic Studies Quarterly*.

Susan Turner Haynes joined Lipscomb University as Assistant Professor in 2015. She earned her doctoral degree and masters degree in political science from George Mason University and her bachelors degree in political science from the University of Tennessee. Prior to graduating from George Mason, Haynes was selected as a Public Policy and Nuclear Threat (PPNT) fellow at the University of California, San Diego. Haynes research specializes in Chinese nuclear strategy and she published her book, *Chinese Nuclear Proliferation*, with the University of Nebraska Press/Potomac Books in 2016. In addition, Haynes' has published numerous articles in academic journals, including in *Asian Perspectives*, *Asian Security*, *PS: Political Science and Politics*, *Strategic Studies Quarterly*, *The Nonproliferation Review*, and *Comparative Strategy*. She has been invited to present her research at the Department of Defense Strategic Multilayer Assessment Group as well as at the Air War College.

Hans M. Kristensen is director of the Nuclear Information Project at the Federation of American Scientists where he provides the public with analysis and background information about the status of nuclear forces and the role of nuclear weapons. He specializes in using the Freedom of Information Act (FOIA) in his research and is a frequent consultant to and is widely referenced in the news media on the role and status of nuclear weapons. Kristensen is co-author of the Nuclear Notebook column in the *Bulletin of the Atomic Scientists* and the *World Nuclear Forces* overview in the Stockholm International Peace Research Institute (SIPRI) Yearbook. The Nuclear Notebook is, according to the publisher, "widely regarded as the most accurate source of information on nuclear weapons and weapons facilities available to the public." Between 2002 and 2005, Kristensen was a consultant to the nuclear program at the Natural Resources Defense Council in Washington, DC, where he researched nuclear weapons issues and wrote the report "U.S. Nuclear Weapons In Europe" (February 2005) and co-authored numerous articles including "What's Behind Bush's Nuclear Cuts" (*Arms Control Today*, October 2004) and "The Protection Paradox" (*Bulletin of the Atomic Scientists,* March/April 2004). Between 1998 and 2002, Kristensen directed the Nuclear Strategy Project at the Nautilus Institute in Berkeley, California, and he was Special Advisor to the Danish Ministry of Defense in 1997–1998 as a member of the Danish Defense Commission. He was Senior Researcher with the Nuclear Information Unit of Greenpeace International in Washington DC from 1991 to 1996, prior to which he coordinated the Greenpeace Nuclear Free Seas Campaign in Denmark, Norway, Finland, and Sweden.

Matthew Kroenig is Associate Professor in the Department of Government and School of Foreign Service at Georgetown University and Deputy Director of the Scowcroft Center for Strategy and Security at the Atlantic Council. In 2019, he was ranked among the top 25 most cited political scientists of his generation. Dr. Kroenig is the author or editor of seven books, including *The Logic of American Nuclear Strategy* (Oxford University Press, 2018). His articles have appeared in many publications, including: *American Political Science Review, Foreign Affairs, Foreign Policy, International Organization, The Wall Street Journal*, and *The Washington Post*. He has served in several positions in the U.S. Department of Defense and the intelligence community and regularly consults with a wide range of U.S. government entities. He has previously worked as a research fellow at the Council on Foreign Relations, Harvard University, and Stanford University. Dr. Kroenig provides regular commentary for major media outlets, including *PBS Newshour, Fareed Zakaria GPS, BBC, CNN, Fox News, NPR,* and *C-SPAN*. He is a life member of the Council on Foreign Relations and holds an MA and PhD in political science from the University of California at Berkeley.

Balazs Martonffy received his doctoral degree from American University's School of International Service in May 2019. His research focuses on international security, alliance cohesion, and NATO. Dr. Martonffy currently is an adjunct instructor at George Washington University's Security Policy Studies Program, and starting in Fall 2019, he will be an assistant professor at the National University of Public Service in Budapest, Hungary. Prior to his doctoral studies, he served as a civilian official in the Hungarian Ministry of Defense. As a Sie Fellow at the University of Denver, he earned a Master's degree in international security in 2012, and as an Eliot Scholar at Washington University in St. Louis, a Bachelor's degree in history in 2010.

Steven Pifer is a William J. Perry fellow at the Center for International Security and Cooperation at Stanford University and a nonresident senior fellow with the Brookings Institution. He focuses on nuclear arms control, Ukraine, and Russia. He has offered commentary on these issues on *National Public Radio, PBS NewsHour, CNN, BBC*, and *VOA*, and his articles have run in *The New York Times, The Washington Post, Financial Times, National Interest, Moscow Times*, and *Kyiv Post*, among others. He is the author of *The Eagle and the Trident: U.S.-Ukraine Relations in Turbulent Times* (Brookings Institution Press, 2017), and co-author with Michael O'Hanlon of *The Opportunity: Next Steps in Reducing Nuclear Arms* (Brookings Institution Press, 2012). A retired Foreign Service officer, his more than 25 years with the State Department included assignments as deputy assistant secretary of state in the Bureau of European and Eurasian Affairs with responsibilities for Russia and Ukraine, ambassador to Ukraine, and special assistant to the president and senior director for Russia, Ukraine, and Eurasia on the National Security Council. He also served at the U.S. embassies in Warsaw, Moscow, and London as well as with the U.S. delegation to the negotiation on intermediate-range nuclear forces in Geneva.

Kingston Reif is the Director for Disarmament and Threat Reduction Policy at the Arms Control Association, where his work focuses on nuclear disarmament, deterrence, and arms control, preventing nuclear terrorism, missile defense, and the defense budget. Reif is an expert on the legislative process and closely monitors Congressional action on these issues. Prior to joining the Arms Control Association, Reif was the director of Nuclear-Nonproliferation at the Center for Arms Control and Nonproliferation and Council for a Livable World. Reif originally came to the Center in 2008 as a Herbert Scoville Jr. Peace Fellow. From September 2008 until May 2009 he served as Dr. Morton Halperin's research assistant on the Congressional Strategic Posture Commission. Reif returned to the Center in May 2009 as Deputy Director of Nuclear Nonproliferation. Reif holds a BA in international relations from Brown University. He spent two years in the U.K. as a British Marshall Scholar where he received a MSc. in international relations from the London School of Economics and Political Science and a MLitt. in international security studies from the University of St. Andrews. He is a 2014 Truman National Security Project Fellow and a 2018 Shawn Brimley Next Generation National Security Leaders Fellow.

Alicia Sanders-Zakre was a research assistant at the Arms Control Association and formerly worked at the Brookings Institution. She graduated from Tufts University *magna cum laude* with a BA in international security.

Cameron Trainer is a research associate at the James Martin Center for Nonproliferation Studies (CNS) of the Middlebury Institute of International Studies at Monterey. He has a background in Russian-language research, covering such diverse subjects as sports policy, military modernization, and sanctions compliance. At CNS, his work primarily focuses on sanctions implementation and compliance. He is also a certified anti-money laundering specialist.

Christian Trotti is a program assistant in the Scowcroft Center for Strategy and Security at the Atlantic Council. In this role, he serves as the Scowcroft Center's lead action officer for U.S. and allied defense issues. Trotti specializes in nuclear deterrence, great power competition, military strategy and operational concepts, the defense industrial base, and wargaming.

Foreword

Traditionally, the policies and strategies that define the potential use or non-use of nuclear weapons, as well as the financial and political costs of maintaining a nuclear force second to none, have been the nearly exclusive purview of government policy-makers, nongovernmental security experts, and academics. Those of us who have spent careers in this field have lamented the absence of a larger civilian audience for these critical issues. With the United States, Russia, and China embarking on substantial and costly nuclear modernization programs, and little prospect in the near term for a return to arms control negotiations, it is time to expand the audience for these important and consequential matters.

A look back at U.S. nuclear history reveals a great deal of continuity and bi-partisanship, although it may not have seemed so when living through it. President Ronald Reagan embraced the vision of a world without nuclear weapons and famously declared in 1984 that a "nuclear war can never be won and must never be fought." With the end of the Cold War came a heightened attention to the challenges of nuclear weapons proliferation. Under successive administrations the U.S. nuclear stockpile grew steadily smaller. The U.S. and Russia continued to pursue negotiated agreements to limit and reduce their nuclear stockpiles, even as national declaratory policies sometimes distracted from these accomplishments.

President Barak Obama's now famous 2009 Prague speech raised expectations that the U.S. was moving to reduce the salience of nuclear weapons in the national security strategy, and to play a lead role in reducing the dangers posed by nuclear weapons by reinforcing the international security architecture through both nuclear arms control agreements and active support for the international nuclear nonproliferation regime, including the Treaty on the Nonproliferation of Nuclear Weapons (NPT). His commitment at that time to pursue a nuclear modernization program to ensure the safety, security and reliability of the U.S. stockpile "as long as nuclear weapons exist," was the "B" side of the nuclear coin but was largely eclipsed by the rest of the speech.

Today the NPT and the global nonproliferation regime is under increasing stress. Frustration with the past pace of nuclear arms control and disarmament has polarized the international community and contributed to the successful negotiation and conclusion of a treaty banning possession and use of nuclear weapons. The collapse of the INF Treaty over unresolved violations, doubts about the extension of the U.S.-Russia New START agreement, and U.S. and Russian modernization programs that would not only produce new nuclear capabilities but also grow their respective arsenals can only widen the chasm between nuclear weapons possessor states and those that have foresworn this capability. This is how arms races are run. And this is why this book is so timely.

The impact of U.S., Russian, and Chinese modernization programs on regional and global stability, alliance relationships and the global nuclear nonproliferation architecture remains to be seen. There is little doubt, however, that modernization programs that reverse what has been a steady decline in the stockpiles of the largest nuclear weapon states over many years, coupled with the absence of bilateral or multilateral arms control initiatives, will severely complicate nonproliferation diplomacy in the foreseeable future and intensify questions about the viability of the NPT.

The editors have collected the contributions of standout global experts who examine and analyze the nuclear modernization programs underway and on the drawing board in the U.S., Russia, and China. *Nuclear Modernization in the 21st Century* explores the motivations for these programs, their consequences for nuclear deterrence, for future arms control initiatives, for the health of the international nuclear nonproliferation regime, and for pursuit of the elusive goal of strategic stability. Readers will better appreciate the debate – and there is a real debate – over the risks, benefits, and financial and political costs associated with these nuclear modernization programs.

Whether one believes nuclear weapons are essential to prevent another Great Power War and must be part of the security equation indefinitely, or that their use under any circumstances is illegal and immoral, those of us who have worked on these issues understand that it is far from a black and white issue. This excellent contribution to the literature and discourse may not change your views, but it will most certainly better inform them.

Ambassador Susan F. Burk
Former Special Representative to the President of the
United States for Nuclear Nonproliferation

Acknowledgments

There are many people I would like to acknowledge in putting this collection together. Firstly, a big thanks goes out to the organizations and the people within them who have supported me in this endeavor, namely: The Australia-U.S. Fulbright Commission, the Arms Control Association, and the Institute for International Science and Technology Policy (IISTP) at George Washington University. Secondly, to my co-editor and colleague, Phil Baxter. We discussed the notion of undertaking a project of sorts way back in 2016 while I was a visiting fellow at the James Martin Center of Nonproliferation Studies – and here we are, book done and dusted. Thanks mate – your work ethic, creativeness, and propensity to get on with the task at hand have been brilliant! To my family who are always one hundred percent behind me on such projects and enable me to pursue my dreams – thank you for everything. Lastly, a massive thanks goes out to the talented and globally recognized academics and analysts who have contributed to this important volume.

Aiden Warren, Melbourne

I would like to thank the contributors to this volume, both those whose work appears in these pages as well as those who intellectually furthered this discussion. I would also like to thank my family for their support during this effort, in particular for the understanding during stretches of writer's block. Finally, I would like to thank Aiden Warren for this unwavering effort to make this volume a reality. His passion for this issue was a constant motivator from initial idea through final edits.

Philip M. Baxter, Washington, DC

Acronyms

ABM Treaty	Anti-Ballistic Missile Treaty
ALCM	Air-launched cruise missile
CEP	Circular error probable
CFE Treaty	Conventional Armed Forces in Europe Treaty
CTBT	Comprehensive Nuclear-Test-Ban Treaty
DOD	U.S. Department of Defense
ICBM	Intercontinental ballistic missile
INF Treaty	Intermediate-Range Nuclear Forces Treaty
IRBM	Intermediate-range ballistic missile
LRSO	Long-range standoff weapon
MIRV	Multiple independently targetable re-entry vehicle
NATO	North Atlantic Treaty Organization
New START	New Strategic Arms Reduction Treaty
NFU	No First Use
NPT	Treaty on the Nonproliferation of Nuclear Weapons
NPR	*Nuclear Posture Review*
PLA	China's People's Liberation Army
SIPRI	Stockholm International Peace Research Institute
SLBM	Submarine-launched ballistic missile
SLCM	Submarine-launched cruise missile
SSBN	U.S. ballistic missile submarine
START	Strategic Arms Reduction Treaty

1 Introduction

Aiden Warren and Philip M. Baxter

Despite the Article VI provision within the 1968 Nuclear Nonproliferation Treaty (NPT) which commits acknowledged nuclear-weapon states to "pursue negotiations in good faith on effective measures relating to cessation of the nuclear arms race at an early date and to nuclear disarmament," there is a renewed interest by several nuclear weapon states in modernizing their arsenals and reaffirming the significance of such weapons. Recent moves and statements indicate substantial investments by nuclear weapon possessor states in the upkeep and modernization of their nuclear postures, signaling a return of primacy and importance of nuclear forces in international politics.

In the United States, policy-makers have called for rapid and significant investment into the nuclear stockpile and strategic systems, which would include new ballistic missile submarines, new strategic bombers, new intercontinental ballistic missiles, enhanced command and control systems, new tactical nuclear systems, and investment into the nuclear weapons complex, in particular, the ability to produce the material for nuclear warheads at a faster rate. Russia, while focusing on legacy systems for much of the 1990s, shifted its resources to the development of new strategic systems in the early 2000s and again in the early 2010s. They have deployed or will likely be deploying new strategic nuclear weapon systems over the next decade. While still possessing a small nuclear weapons arsenal in comparison to the United States or Russia, a shift in Chinese nuclear strategy has been paired with a rapidly growing and diversifying nuclear arsenal.

With these developments in mind, it could be argued that a realist outlook and deterrence models appear to be taking precedence over those that advocate nuclear arms control and disarmament.[1] The pace of nuclear reductions has slowed markedly with Russia and the United States clearly shifting their emphasis to sustaining or expanding their arsenals for the indefinite future. Additionally, with the deterioration of bilateral relations between Moscow and Washington impeding any new arms control agreements, no multilateral disarmament negotiations in sight, and a clear preference towards nuclear modernization, some observers argue whether the nuclear nonproliferation regime is in crisis, and question its long-term prospects. China, India, North Korea, Pakistan, and Israel are emboldening their stockpiles, signifying the extent to which all nuclear-armed states appear to be viewing nuclear weapons as an enduring and indefinite mainstay of their respective national and international security calculi.

An additional factor in international politics is that while the perceived value of nuclear weapons appears to have shifted (from diminishing to expanding) for nuclear weapon possessors, most other states recently committed to a global ban on these weapons, increasing the tension on the continued viability of the NPT. It could be argued that the salience of nuclear weapons and great power competition has not been at current levels since the end of the Cold War. The rapid build-up and diversification of nuclear weapon arsenals introduce potential security challenges that have not been seen since the 1980s, outside of the vertical proliferation associated with the Pakistan-India competitive dyad and the challenges from the emergence of North Korea as a nuclear-weapon state. Notwithstanding the economic restrictions facing several of the nuclear-armed states, the United States, Russia, and China, among others, appear committed to spending hundreds of billions of dollars over the next decade on modernizing their nuclear forces. Indeed, while the nuclear arms race between East and West concluded nearly 30 years ago, a vigorous technological nuclear arms race is gathering momentum and will continue to escalate over the next several decades as new systems are placed online and states respond to match the deployment.

In providing a substantive analysis of this emerging security development, this volume seeks to advance policy debate in ways that support international security, while also filling gaps in the academic analysis of nuclear weapons and nonproliferation. Moreover, in providing a review of the nuclear modernization across the technical, policy and strategic domains, this collection attempts to add important connective tissue in bridging the various scholarly literature, and overall, providing a comprehensive assessment of the impact of nuclear modernization on broader international security. As the volume concerns the United States, Russia, and China – the three most defining state actors in the nuclear weapon domain and international security – we feel that unpacking their respective modernization drives is both timely and necessary.

The remainder of this introductory chapter will proceed in the following order. First, it is important to acknowledge that modernization debates remain a source of consternation and look to further intensify given the position of power politics and the emergence of new technologies. With this in mind, we then proceed to provide what we think is critical to the book's foundation: defining the focus of our attention, nuclear modernization. Next, we set the stage with a brief summary of recent developments in the nuclear weapons domain for each of the three countries to orient the reader. Finally, we outline the plan for the remainder of this volume with a discussion of each chapter's contribution to the analysis.

Modernization debates

The debates pertaining to nuclear modernization are wide and varied. Proponents of modernization, particularly those in the United States disagree that any notion of nuclear disarmament requires the United States to cut its arsenal and limit its modernization drive. They argue that the United States has already significantly reduced its stockpile over the years, while conversely, other states have moved in the opposite direction, expanding their nuclear forces. While they acknowledge comprehensive nuclear disarmament may be an appealing and laudable goal, accomplishing it will need to entail a massive transformation of the international political and security environment. In this regard, "failure to modernize would not contribute to disarmament – but more than that, it would be irresponsible. [it] would embolden enemies, frighten allies, generate international instability, and undermine U.S. national security."[2]

Other proponents of modernization argue that if the United States does not maintain a credible deterrent, there could be "painful and possibly disastrous consequences for U.S. interests in the world." As such, more agility in the form of weapons that can be used in smaller-scale, conventional conflicts are needed, including the low-yield, tactical nuclear weapons.[3] The 'flexible' and 'agile' terms are central concepts to what proponents of modernization see as being imperative in addressing threats in a rapidly changing world. Here, they contend, "the nuclear weapons enterprise must be flexible and resilient to underpin the U.S. nuclear deterrent."[4] In this context, if the United States detects a significant nuclear weapons development drive in another state, its nuclear weapons complex must have the capacity, and 'agility', to respond to such threats in a timely fashion. It is not surprising that many modernization proponents fall in the nuclear deterrence camp; a camp where it has long been argued that U.S. nuclear forces play a defining role in the global nonproliferation regime by providing security guarantees and assurances to NATO, Japan, and South Korea. In providing such deterrence mechanisms, allies either keep the numbers of their nuclear weapons relatively low (France and the United Kingdom) or relinquish their 'opportunity' to develop and deploy such weapons altogether. Should the integrity of American nuclear forces continue to diminish, states like South Korea may pursue their own nuclear option, which would engender varied complications to broader security across the region.[5]

At the moderate level, an acknowledgment to the dramatic change of the security environment since 2010 has started to be included in recent assessments. Most nuclear weapon states, including the United States and Russia, have often professed their commitment to reducing their respective nuclear weapons stockpiles. However, while this may be the case, it is evident that no nuclear weapon state is prepared to wholly relinquish its nuclear capabilities unless all other states do likewise. As such, nuclear arms are likely to remain in military arsenals for an extended period of time and will continue to be refurbished. While one may accept nuclear modernization as being unavoidable, the question persists as to the extent to which it can be balanced in such a way that it does not create impediments to nuclear arms reductions and to complete disarmament in the long run.[6] In pointing to the seminal Barack Obama speech in Prague in April 2009 as a point of optimism, Frank Rose and Benjamin Bahney argue that instead of joining the United States in strengthening efforts to alleviate nuclear threats, Russia and China have moved in the opposite direction through investing in new nuclear weapons systems, conventional strike, and asymmetric capabilities.[7] As indicated, given the intensification of power politics since Obama's speech, some analysts and policymakers in the moderate camp have adjusted their thinking, arguing that amid such global security realities, it is now critical to modernize the strategic nuclear deterrent in a way that reassures allies and augments strategic stability.[8] That said, in perfect hedging fashion, they argue strategic nuclear modernization alone is not ample to maintain the security of the United States and its allies, and as such, arms control and nonproliferation need to be incorporated into the strategic equation.[9]

For Lu Yin, while eliminating nuclear weapons does not appear possible at this stage, it is important to attain a balance between modernization and disarmament. This approach, however, would require nuclear strategy to be adjusted so that nuclear weapons hold a diminished role in a states' national security, and thereby, the practical reasons for even possessing nuclear weapons would wane. Of course, this would be very much dependent on the United States and Russia, the two nuclear superpowers, taking a leading role, establishing trust, and setting a good example for other states.[10] Given the current global security context, even the most forthright of optimists would struggle to see the possibility of such 'example setting'; particularly when viewing the current trajectory of nuclear modernization programs. In terms of the United States, according to Andrew Weber and Christine Parthemore, the U.S. modernization drive departs markedly in several ways from longstanding U.S. nuclear policy. Simply put, it has amplified the role of nuclear weapons in U.S. national security strategy, encompassing "plans for developing several new nuclear weapon capabilities, and resurrects former nuclear capabilities that past U.S. presidents had wisely eliminated." In recent times, the Trump administration's nuclear modernization plans that include low-yield nuclear options and dual-capable systems that may be abstrusely nuclear or conventional, have intensified debates across the political spectrum.[11]

A recent global evaluation on the present state of armaments, disarmament and international security, has argued that notwithstanding the reductions made to their respective nuclear forces, both Russia and the United States have long-term programs underway to replace and modernize their nuclear warheads, missile and aircraft delivery systems, and nuclear weapon production facilities. The most defining aspect of the report was the preparedness of both states to not just modernize for safety and assurance purposes, but to develop new nuclear weapons and expand nuclear options to deter and, if necessary, defeat both nuclear and non-nuclear strategic attacks.[12] According to Ambassador Jan Eliasson, Chair of the Stockholm International Peace Research Institute (SIPRI) Governing Board, "the renewed focus on the strategic importance of nuclear deterrence and capacity is a very worrying trend,"[13] while a report by the Arms Control Association has argued that nuclear modernization is "unnecessary, unsustainable, and unsafe" and clearly increases "the risks of miscalculation and unintended escalation, and accelerated global nuclear competition."[14]

A more global perspective illustrates the gravity of the situation. With all nuclear-armed states modernizing and upgrading their arsenals, many analysts and policymakers believe that a new nuclear arms race is underway. However, this version will be different and will encompass greater security concerns based on it involving three competing great powers rather than two. This will of course, be exacerbated with the advent of new non-nuclear weapon technologies such as ballistic missile defense, anti-satellite weapons, precision-strike missile technology and artificial intelligence (AI), making nuclear deterrence relationships more uncertain and broader global security more unstable.[15]

Defining modernization

Given the varied and wide dichotomies between many analysts, academics and policy makers on the nuclear modernization debate, it is imperative that we, the editors, posit our working standpoint. As such, before proceeding into a discussion of the outline of this book, it is important to clearly elucidate the terms that will be used throughout this book, specifically how *we* conceptualize "modernization."

In terms of this book, we take on the following position, where modernization consists of three elements of nuclear weapon capabilities. The first resides in the context of aging arsenals and subsequent efforts to construct new, but roughly similar, weapons with the goal of retiring or supplementing existing stockpiles. Related, the second pillar consists of the modernization and expansion of the size of nuclear arsenals. Finally, modernization includes the development of new types of nuclear weapons, delivery platforms, or integration of new technologies into strategic doctrine to supplement the nuclear arsenal.

It should be noted, however, to limit modernization to only one segment would miss or discount issue linkages and security implications. For example, if we limit the modernization discussion to only swapping old weapons for more up-to-date ones, the developments in arsenal expansion, technology advancement, and technology integration would be left unexamined. Along the same line, if we touch only on proposed technological advancements or integration with other weapon platforms, our examination would not address the salient issue of arsenal fatigue. With this in mind, while this volume takes a wide lens approach to defining modernization, this is necessary in order to address the shifts in nuclear weapon arsenals in the United States, China, and Russia.

Finally, our working definition of modernization goes beyond simply the nuclear warheads themselves. Modernization efforts apply to all elements of strategic weapons – to include the various delivery systems. While pit production capability and warhead fabrication capabilities are important to the discussion, how a state chooses to invest to deploy its nuclear weapons is equally important as this can determine its deterrent strategy and strategic objectives.

The following discussion includes all three elements under the banner of modernization as the perspectives and strategic intent of nuclear modernization efforts by countries are diverse.

Setting the stage

During the Barack Obama presidency, the administration embarked on an overhaul of its entire nuclear weapons enterprise, encompassing the development of new weapons delivery systems and modernizing its enduring nuclear warhead types and nuclear weapons production facilities in a program that scholars estimate could cost more than a trillion dollars.[16] The modernized B61-12 warhead, for example, will be able to strike targets more accurately with a smaller explosive yield and reduce the radioactive fallout from a nuclear attack, and could thus become more militarily attractive and increase the likelihood of use. Other modifications under consideration, such as interoperable warheads that could be used on land- and sea-based ballistic missiles, could significantly alter the structure of the nuclear warheads and potentially introduce new uncertainties into the global nuclear order.[17] Further, with the election of Donald Trump, it is evident that the nuclear option encompassing new weapons to replace the current ones will – most concerning – introduce new military capabilities to the weapons system. The new *Nuclear Posture Review* (NPR), compounded with President Trump's disdain for arms control treaties, as evidenced by the U.S. withdrawal from the Intermediate-Range Nuclear Forces (INF) Treaty on August 2, 2019, has the potential to negatively impact the positive role that the United States can and must play in moving toward a world without nuclear weapons and sustaining the NPT.

Since the collapse of the Soviet Union, nuclear weapons have largely underwritten Russian national security. During this period, Russia's conventional military power waned and still remains lacking in comparison to other nations. This has put the focus of investment upon strategic weapons and delivery systems through much of the period since the early 2000s (after an initial investment into Cold War-era legacy systems in the early 1990s), as well as adopting strategies of asymmetric means of conflict. During the 1990s, the scale of the arsenal was scaled back, due to treaty obligations, loss of production centers for components when the Soviet Union dissolved, and financial constraints. This put the focus of investment in maintaining center key weapons systems.

In the early 2000s, a renewed focus was placed on the need to modernize its nuclear arsenal, a move many have attributed to the actions and high-tech demonstration by the United States in Iraq and Afghanistan. These efforts have included upgrading an intercontinental ballistic missile to carry additional warheads, developing a replacement intercontinental ballistic missile (ICBM) to replace its mobile-launch ICBM, developing a new class of nuclear submarines and sea-launched ballistic missiles, developing a new heavy bomber, advancing dual-use hypersonic weapons research and testing, a nuclear-powered cruise missile, and a nuclear-powered underwater drone. Moreover, the efforts demonstrate the continued importance that the nuclear arsenal and different legs of the nuclear triad play in Russia's security policies.

In the context of the Chinese nuclear arsenal, it has been undergoing a series of modernization efforts since the 1980s, primarily targeted at upgrading its land-based missile force.[18] These missile improvements have included the development and deployment over the last few years of two new road-mobile intermediate-range ballistic missiles and one new intercontinental ballistic missile launcher.[19] China has also deployed it's first multiple independently targetable re-entry vehicle (MIRVed) missile and is working to complete a new road-mobile ICBM.[20]

The modernization of its submarine force is also a high priority, with plans to expand its 60 submarine fleet, including its current four nuclear-powered ballistic missile submarines.[21] According to the U.S. Department of Defense, these "four operational JIN-class U.S. ballistic missile submarines (SSBNs) represent China's first credible, sea-based nuclear deterrent."[22] China is expanding its fleet of long-range bombers, a variant of the H-6 BADGER, which is capable of dual nuclear-conventional missions, although it does not currently carry a primary nuclear mission.[23] Under the current modernization efforts, the Chinese nuclear forces are slowly increasing in the quantity of its nuclear arsenal, as well as the quality.

The policy significance of this effort rests on the insights provided by some of the world's leading analysts and academics into how international nuclear modernization will have an impact on disarmament, bilateral arms control, nuclear security, and the overall mobilization of political will in other states. The chapters will inform scholars, analysts, academics and policy-makers by connecting analytical and normative insights that can inform policy decisions to simultaneously support the international security interest in stabilizing the NPT and progressing disarmament, while also supporting the legitimate security interests of the United States and its allies with a safe and strategically reassuring nuclear deterrent.

Plan of the book

The book proceeds across three themes examining the technical developments in the United States, China, and Russia regarding nuclear weapons, the policy implications of these developments, and the strategic outlook from modernization. This approach was selected as nuclear weapons modernization operates at the intersection of these components, and as stated above, all impact disarmament, bilateral arms control, nuclear security, and the overall mobilization of political will in other states. The early stages of the book consist of chapters which will each in-turn examine the U.S., China, and Russia and discuss developments within the strategic nuclear weapon arena. These chapters will highlight the technical developments, planned or proposed, for the strategic systems of the three countries.

In Chapter 2, Hans Kristensen examines the United States' nuclear weapons modernization efforts and the 2018 *Nuclear Posture Review*. The chapter reviews the upgrades planned in air-delivered nuclear weapons, land-based ballistic missiles, ballistic missile submarines, non-strategic nuclear weapons, and the nuclear weapons production complex. Additionally, Kristensen examines the impact of modernization programs on the U.S. nuclear posture. His assessment of the 2018 *Nuclear Posture Review* maintains that the document continues the large-scale modernization efforts begun under the Obama administration, with several modifications. In particular, the document shifts away from aspirations of reductions in nuclear weapons, to a more assertive posture focused on great power competition.

In Chapter 3, Philip M. Baxter reviews developments in Russian strategic weapons, examining the transition from the Cold War era to the present. The chapter examines the different weapons platforms and their evolution, while also attempting to situate this evolution in technical development within the context of the motivating factors behind these developments. The chapter looks at legacy systems, new developments, and why these developments matter, in particular within the context of U.S.-Russia strategic parity. The chapter highlights two persistent themes throughout the three decades since the fall of the Soviet Union: the continued necessity and value which strategic nuclear arms plays in Russia, and their role in domestic politics.

Utilizing pertinent state case studies, the book will attempt to contextualize modernization efforts within both changes to strategic nuclear policies as well as impact on arms control policies. In Chapter 4, Cameron Trainer examines the action-reaction dynamics of Russian and American nuclear modernization through three lenses: strategic, non-strategic, and arms control. The strategic nuclear weapons discussion centers on the classic components of the nuclear triad, ground-, air-, and sea-launched delivery systems, as well as the role that evolving interests and developments in missile defense programs have had on the strategic outlook. The non-strategic nuclear weapons portion of the chapter reviews the ways that the United States and Russia envision using them, in particular focusing on debates surrounding the Russian 'escalate to de-escalate' doctrine section. The final portion of the chapter will assess the impact of evolving developments and perceptions regarding modernization in the two countries on future arms limitation agreements between the two states.

In Chapter 5, Susan Turner Haynes surveys the evolution of Chinese nuclear strategy as the country's nuclear arsenal has rapidly grown and diversified. She examines increasing evidence indicating that China is pursuing a transition from a minimum deterrent capability, a posture with relatively few options for nuclear response, to full spectrum deterrence, a more flexible deterrence posture which would allow China to engage in nuclear war-fighting. The chapter proceeds in three sections: a clarification of concepts, a review of primary source material on China's strategic transition, and a forecast of China's nuclear weapon outlook for the future.

In Chapter 6, Aiden Warren examines the tensions between U.S. nuclear force modernization and the global nonproliferation regime, particularly in the context of weakening U.S. arms control/disarmament leadership and an increasingly challenging international security environment. While some modernization efforts are already underway, there has been extensive debate surrounding the massive investments that such modernization will require, the extent to which new 'flexible' modes will impact on broader global security and the strategic calculi of other states, and as this chapter will discuss, the ramifications such modernization will have on the United States in meeting its obligations under the NPT. In unpacking and assessing this debate, the chapter will commence by providing an overview of the NPT, including the recent strains it has been facing in the context of the NPT review conference process. The chapter proceeds to assess and elucidate the role that the United States has played in driving the NPT and broader regime through its leadership and political will, albeit that has wavered at various intervals since its inception in 1968. Lastly, the chapter will look specifically at U.S. nuclear modernization and examine the tensions between U.S. nuclear modernization programs and the divisions this is producing within the NPT. Given the growing frustrations of non-nuclear weapon states over the slow pace of disarmament and a deteriorating U.S.-Russian relationship that threatens existing arms control treaties, the chapter examines the role that nuclear modernization programs play in exacerbating these fractures.

In Chapter 7, Steven Pifer examines the North Atlantic Treaty Organization's modernization efforts against the backdrop of a European security environment that has changed dramatically and for the worse since the 1990s, particularly over the past five years. With that in mind, he suggests that as NATO proceeds with its nuclear modernization, the alliance should consider how to enhance deterrence while also seeking to contain and manage tensions with Russia. NATO should structure its nuclear posture so as not to lower the threshold for use of nuclear weapons in a conventional conflict. Properly designed arms control could contribute to NATO security but appears unrealistic at this point. The more immediate challenge is how to deal with the fraying U.S.-Russia nuclear arms control regime.

Transitioning from the policy-focused section to one more aimed at theoretical and strategic discussions, Chapter 8 provides a theoretical overview and assessment by Balazs Martonffy and Eleni Ekmektsioglou. They first provide an overview of the main Cold War theories and the way they conceptualized the impact of nuclear weapons on state interactions. Next, they examine how current modernization efforts, in particular by the U.S., might influence traditional theories of international relations and consider the role of middle-range theories. The last section looks at some critical foreign policy questions such as the U.S.-China relationship and the impact of U.S. modernization on theories within the nuclear nonproliferation literature.

In Chapter 9, Kingston Reif and Alicia Sanders-Zakre examine the ways in which the Trump administration's nuclear strategy is unnecessary, unsustainable, and unsafe, and overall, can be considered a determent to international security. It describes three realistic options to reduce spending on nuclear weapons while still maintaining a devastating nuclear deterrent, where scores of billions of dollars could be saved or redirected to higher priorities by eliminating, delaying, or scaling back the administration's proposals for new delivery systems, warheads, and infrastructure. Despite claims that nuclear weapons "don't actually cost that much," the simple fact is that unless the administration and its successors find a 'pot of gold at the end of the rainbow,' planned spending to maintain and replace the arsenal will pose a significant affordability problem, and threaten other national security priorities.

Moreover, the plans appear likely to increase the risks of miscalculation and unintended escalation, undermine strategic stability, and accelerate global nuclear competition. Reif and Sanders-Zakre conclude by arguing that now is the time to re-evaluate nuclear weapons spending plans before the largest investments are made. The choice then is between: the current strategy, which is excessive and unnecessary, puts the United States on course for a budgetary 'train wreck,' and could increase nuclear risk; or a more realistic and affordable approach that still leaves the United States with a devastating nuclear force that is more than capable of deterring any nuclear threats to the United States and its allies.

In Chapter 10, Matthew Kroenig and Christian Trotti argue that modernization programs of the United States and its democratic allies at the core of the international system tend to promote international security, while conversely, the nuclear modernization programs of revisionist autocratic powers intent on challenging the international system tend to threaten global peace and stability. Because all modernization programs are not created equally, it is thereby the underlying geopolitical conditions, and not solely the capabilities themselves, which determine the extent to which modernization programs promote or diminish international security. In more straightforward terms they argue, geopolitical intentions guide nuclear strategy, which in turn informs modernization programs. It is, therefore, essential to account for a state's position within the international system when evaluating the likely effects of nuclear modernization on international security. The remainder of the authors' chapter proceeds in four parts. First, it explains why U.S. nuclear weapons serve as a unique guarantor of international security in the postwar world order. Second, it argues that the United States must modernize its aging nuclear capabilities in order to sustain its longstanding role in promoting global security. Third, it addresses the ways in which contemporary revisionist powers seek to use their nuclear modernization programs for destabilizing ends. Fourth, and finally, the chapter concludes with policy implications.

Rather than a summarization of the preceding chapters, the book concludes with a forward-looking review of scenarios and futures which potentially could arise within three distinct domains given the trajectory of nuclear modernization and nuclear weapons outlook discussed over the course of the volume. These critically impacted domains include strategic stability, great power conflict, and arms control and nonproliferation. Finally, brief closing thoughts on recommendations and potential pathways for reducing potential conflict and misinterpretation of nuclear modernization efforts are provided.

Notes

1 Kingston Reif, "U.S. Nuclear Modernization Programs," *Arms Control Association*, August 2018, www.armscontrol.org/factsheets/USNuclearModernization.
2 Matthew Kroenig, "Why U.S. Nuclear Modernization Is Necessary," *Bulletin of Atomic Scientists*, January 8, 2015, https://thebulletin.org/roundtable/modernizing-nuclear-arsenals-whether-and-how/.
3 Lara Seligman, "Will Congress Let Trump Build More Nuclear Weapons? The Administration and Capitol Hill Are on a Collision Course Over the Future of U.S. Nukes," *Foreign Policy*, April 2019, https://foreignpolicy.com/2019/04/11/will-congress-let-trump-expand-americas-nuclear-arsenal/.
4 "U.S. Nuclear Weapons Capability," October 4, 2018, www.heritage.org/military-strength/assessment-us-military-power/us-nuclear-weapons-capability.
5 Ibid.
6 Eugene Miasnikov, "Modernization and 'Zero': Compatible Tendencies?" *Bulletin of Atomic Scientists*, January 8, 2015, https://thebulletin.org/roundtable/modernizing-nuclear-arsenals-whether-and-how/.
7 Frank A. Rose and Benjamin Bahney, "Reassuring Allies and Strengthening Strategic Stability: An Approach to Nuclear Modernization for Democrats," *War on the Rocks*, April 16, 2019, https://warontherocks.com/2019/04/reassuring-allies-and-strengthening-strategic-stability-an-approach-to-nuclear-modernization-for-democrats/.

8 Ibid.
9 Ibid.
10 Lu Yin, "Balancing Modernization and Disarmament," *Bulletin of Atomic Scientists*, January 8, 2015, https://thebulletin.org/roundtable/modernizing-nuclear-arsenals-whether-and-how/.
11 Andrew Weber and Christine Parthemore, "Smarter U.S. Modernization, without New Nuclear Weapons," *Bulletin of Atomic Scientist*, volume 75, 2019, 25–29.
12 "World Nuclear Forces: Reductions Remain Slow as Modernization Continues," *The Stockholm International Peace Research Institute (SIPRI)*, June 18, 2018, www.sipri.org/media/press-release/2018/modernization-nuclear-weapons-continues-number-peacekeepers-declines-new-sipri-yearbook-out-now.
13 Ibid.
14 Kingston Reif and Alicia Alicia Sanders-Zakre, "U.S. Nuclear Excess: Understanding the Costs, Risks, and Alternatives," *Arms Control Association*, April 2019, www.armscontrol.org/reports/2019/USnuclearexcess.
15 Ben Zala, "How the Next Nuclear Arms Race Will Be Different from the Last One," *Bulletin of the Atomic Scientists* 75, no. 1 (2019), 36.
16 Jon B. Wolfsthal, Jeffrey Lewis, and Marc Quint, "The Trillion Dollar Nuclear Triad," *James Martin Center for Non-Proliferation Studies*, January 2014; Hans M. Kristensen, "Nuclear Weapons Modernization: A Threat to the NPT," *Arms Control Today*, 2014; Reif, "U.S. Nuclear Modernization Programs"; William J. Broad, and David E. Sanger, "U.S. Ramping Up Major Renewal in Nuclear Arms," *The New York Times*, Section A, Page 1, September 12, 2014.
17 Kristensen, "Nuclear Weapons Modernization: A Threat to the NPT."
18 Hans M. Kristensen and Robert S. Norris, "Chinese Nuclear Forces, 2018," *Bulletin of the Atomic Scientists* 74, no. 4 (2018), 290–291.
19 Ibid., 290.
20 U.S. Department of Defense, "Military and Security Developments Involving the People's Republic of China," April 2015, www.defense.gov.
21 Office of the Secretary of Defense, "Annual Report to Congress: Military and Security Developments Involving the People's Republic of China 2019," May 2, 2019, 35–56.
22 Ibid., 36.
23 Office of the Secretary of Defense, "Annual Report to Congress," 41; Kristensen and Norris, "Chinese Nuclear Forces, 2018," 293.

2 U.S. modernization efforts and the 2018 *Nuclear Posture Review*

Hans M. Kristensen

The United States has reduced its nuclear arsenal significantly since the end of the Cold War, both in terms of numbers of warheads and their delivery vehicles. The Department of Defense is thought to maintain a stockpile of an estimated 3,800 nuclear warheads for delivery by more than 800 ballistic missiles and aircraft. Most of the warheads in the stockpile are not deployed, however, but stored for potential upload onto missiles and aircraft if necessary. Many are destined for retirement. Roughly 1,750 warheads are currently deployed, of which about 1,300 strategic warheads are deployed on ballistic missiles, 300 at bomber bases in the United States, and another 150 non-strategic bombs are deployed at air bases in Europe. The remaining warheads – approximately 2,050 – are in storage as a so-called hedge against technical or geopolitical surprises. Several hundred of those warheads are scheduled to be retired before 2030. In addition to the warheads in the Department of Defense stockpile, approximately 2,340 retired, but still intact, warheads are stored under the custody of the Department of Energy and are awaiting dismantlement, giving a total U.S. inventory of an estimated 6,140 warheads.[1]

The remaining weapons, however, are in the first phase of the most comprehensive modernization of the U.S. nuclear arsenal since the end of the Cold War. The effort is scheduled to last well into the 2040s and cost, together with the price of operating existing forces, at least $1.5 trillion.[2] The core of the modernization plan – known as the Program of Record – is designed to renew the entire triad of strategic nuclear forces, tactical nuclear weapons, nuclear command and control, nuclear warheads, and the infrastructure that builds them.

The plan was drawn up and started during the Obama administration. Since then, relations with Russia and China have deteriorated significantly and the Trump administration has taken office with a pledge to strengthen U.S. nuclear capabilities and denouncing arms control agreements. The hawkish new administration has embraced what it coins the return of Great Power competition, publicly criticized its Allies, cozied up to authoritarian regimes and publicly questioned the U.S. intelligence community's conclusion that several adversaries are conducting cyber-attacks against the United States and its Allies, withdrawn the United States from landmark nuclear agreements such as the Iran Agreement (Joint Comprehensive Plan of Action; JCPOA) and the Intermediate-Range Nuclear Forces Treaty (INF), and questioned the need to extend the New START treaty.

The Trump administration's National Security Strategy, Nuclear Posture Review, and its latest Stockpile Stewardship and Management Plan contain much continuity but also breaks with earlier U.S. nuclear policy by proposing new nuclear weapons, enhancing the role of nuclear weapons, and reducing the emphasis on arms control.

This chapter provides a brief overview of the various nuclear weapons modernization programs and their impact on the U.S. nuclear posture. For each of the weapons categories described below, the modernization program also includes extensive upgrades planned for the nuclear command and control (NC3) systems that are used to manage and operate the arsenal, plan the strike plans, and employ the weapons if so ordered by the National Command Authority. But before outlining the programs, a brief review of the *Nuclear Posture Review* is in order to put things into context with the current political situation.

The 2018 *Nuclear Posture Review*

The *Nuclear Posture Review* (NPR) published by the Trump administration in February 2018 is the fourth NPR conducted since the end of the Cold War. The first was the Clinton administration's NPR in 1994, followed by the W. Bush administration's review in 2001, and the Obama administration's review in 2010. Despite the Trump administration's bombastic rhetoric about building up nuclear forces, the main outline of its NPR is focused on carrying forward the large-scale modernization program started by the Obama administration.

The NPR belittles the projected cost of the modernization program and claims it is "an affordable priority" and emphasizes that the total cost is only a small portion of the overall defense budget.[3] There is little doubt, however, that limited resources, competing nuclear and conventional modernization programs, tax cuts, and the rapidly growing deficit will present significant challenges for the nuclear modernization program.

Despite similarities, the Trump NPR differs from the previous administration's NPR from 2010 on several key issues. The most significant change is a shift away from seeking to reduce the number and role of nuclear weapons to a more confrontational tone and assertive posture that embraces so-called Great Power competition. The NPR also breaks with previous policy about not building new nuclear weapons and proposes "nuclear supplements" to the arsenal. Moreover, the NPR moves away from the 2010 NPR's goal of seeking to limit the role of nuclear weapons to the sole purpose of deterring nuclear attacks, and instead emphasizes "expanding" U.S. nuclear options to deter and, if deterrence fails, defeat both nuclear and "non-nuclear strategic attacks."[4]

Nuclear attacks against non-nuclear strategic attacks are first-use scenarios. The NPR defines that "non-nuclear strategic attacks include, but are not limited to, attacks on the U.S., allied, or partner civilian population or infrastructure, and attacks on U.S. or allied nuclear forces, their command and control, or warning and attack assessment capabilities."[5] U.S. nuclear capabilities will be postured to "hedge against the potential rapid growth or emergence of nuclear and non-nuclear strategic threats, including chemical, biological, cyber, and large-scale conventional aggression."[6] To achieve this aim, the NPR states, "the United States will enhance the flexibility and range of its tailored deterrence options. . . . Expanding flexible U.S. nuclear options now, to include low-yield options, is important for the preservation of credible deterrence against regional aggression."[7]

The new tailored capabilities include, in the short term, the modification of "a small number" of W76–1 warheads on the Trident II D5LE submarine-launched ballistic missile (SLBM) "to ensure a prompt response option that is able to penetrate adversary defenses."[8] The National Nuclear Security Administration's (NNSA's) latest Stockpile Stewardship and Management Plan (SSMP) describes the requirement this way: "The W76–2 provides a low-yield option capable of overcoming adversary air defenses."[9] This capability is necessary, the NPR claims without providing evidence why, to "help counter any mistaken perception of an exploitable 'gap' in U.S. regional deterrence capabilities."[10] This part of the plan modifies some of the existing 100-kiloton W76–1 warheads to the low-yield (possibly 5–7 kilotons) W76–2 warheads. The first W76–2 was completed in February 2019 and will be delivered to the Navy in 2020, although it remains to be seen if Congress will agree to pay for the deployment.

In the longer term, the NPR presents a plan to "pursue a nuclear-armed" submarine-launched cruise missile (SLCM) that will also have low-yield options. The Navy had a nuclear SLCM in its inventory until 2011, when it was retired by the 2010 NPR because it was redundant.[11] The 2018 NPR recommends acquiring a nuclear SLCM again to "provide a needed non-strategic regional presence, an assured response capability," and a treaty-compliant response to Russia's violation of the INF treaty.[12] This last objective is now moot because the Trump administration has pulled out of the INF treaty. The NPR states that the U.S. "will immediately begin efforts to restore this [SLCM] capability by initiating a requirements study leading to an Analysis of Alternatives . . . for the rapid development of a modern SLCM."[13] As further stated,

> . . . the U.S. pursuit of a submarine-launched cruise missile may provide the necessary incentive for Russia to negotiate seriously a reduction of its non-strategic nuclear weapons, just as the prior Western deployment of intermediate-range nuclear forces in Europe led to the 1987 INF Treaty.[14]

These short- and long-term nuclear "supplements" will, the NPR explains,

> provide a more diverse set of characteristics greatly enhancing. . . [the U.S.'s] ability to tailor deterrence and assurance; expand the range of credible options for responding to nuclear or non-nuclear strategic attack; and enhance deterrence by signaling to potential adversaries that their concepts of coercive, limited nuclear escalation offer no exploitable advantage.[15]

The claim that new low-yield nuclear capabilities are needed is somewhat of a mystery because the U.S. nuclear arsenal already includes around 1,000 gravity bombs and air-launched cruise missiles (ALCMs) that have low-yield warhead options.[16]

The new nuclear weapons are primarily being justified by pointing to Russian non-strategic nuclear weapons, which are being modernized. Russia has had a much larger inventory of such weapons than the United States for decades, but that has not prevented the U.S. military from retiring most of its non-strategic weapons; there was no longer a military need for them. Russia developed its new INF-violating SSC-8 GLCM in 2008–2010 at a time when the U.S. arsenal included a nuclear SLCM, so it is therefore unclear why the NPR now states that reintroduction of a nuclear SLCM would cause Russia to change its behavior. Moreover, U.S. Strategic Command has already strengthened U.S. strategic bomber support of the North Atlantic Treaty Organization (NATO) to enhance deterrence in response to the perceived provocative and aggressive behavior of Russia.[17] Those bombers are equipped with the nuclear ALCM and gravity bombs (in the case of the B-2) and will receive the new air-launched cruise missile (the LRSO or long-range standoff missile), which will have essentially the same capabilities as the new SLCM proposed by the NPR (see later).

Moreover, a new SLCM would require installation of nuclear-certified storage and launch-control equipment on the attack submarines that are assigned the new mission. Sea- and land-based personnel would need to be trained and certified to maintain and handle the weapons. These are complex and expensive logistical requirements that would place further strain on the U.S. Navy's financial and operational resources. In addition, the fielding of a nuclear SLCM could spark discord with certain states because public protests against nuclear-capable warship visits could flare-up.[18] In the case of New Zealand, for example, that country's long-standing policy of refusing entry to its ports of nuclear-capable vessels collided with the U.S. policy of refusing to disclose – directly or indirectly – whether its vessels carried nuclear weapons and created significant strains in its diplomatic relations with the U.S. in the mid-1980s; strains that have only recently begun to be repaired.[19] The U.S.'s reintroduction of a nuclear SLCM could reopen such political disputes and complicate relations with key allied countries in Europe and North East Asia. These additional costs should be weighed against the benefits that the NPR claims a new SLCM would provide.

The NPR's strategy to counter Russian non-strategic nuclear weapons with more non-strategic nuclear weapons is a misguided and risky gamble. The recommendation to acquire two new nuclear weapons to "help counter any mistaken perception of an exploitable 'gap' in U.S. regional deterrence capabilities" appears to communicate to Russia and U.S. NATO allies that the Trump administration does not consider its current deployment of dual-capable aircraft and bombs in Europe to be a credible deterrent.

Moreover, deploying low-yield warheads on fast-flying ballistic missiles on strategic submarines for use early in a conflict in response to Russian non-strategic nuclear weapons not only blurs the line between strategic and non-strategic nuclear weapons, but risks unintentionally escalating a conflict to use of strategic nuclear weapons. Moscow would misinterpret use of a low-yield W76–2 fired from a

Trident submarine as the start of a U.S. strategic attack on Russian territory, not a limited attack intended to de-escalate a conflict.

It is also misguided that Russia's decisions about the size, composition, and role of its non-strategic nuclear arsenal are guided by the U.S. non-strategic nuclear arsenal or what yield it has. Rather, it appears to be driven primarily by the U.S. superior *conventional* forces.[20] Instead, pursuit of a new SLCM to "provide a needed non-strategic regional presence" in Europe and Asia could – especially when combined with the parallel expansion of U.S. long-range conventional strike capabilities – deepen rather than reduce Russia's reliance on non-strategic nuclear weapons and potentially could even trigger Chinese interest in enhancing its shorter-range weapons as well.

Apparently well aware that new low-yield nuclear weapons would spark accusations that the Trump administration is increasing U.S. nuclear war-fighting capabilities, the authors of the NPR insisted that the new weapons will not lower the threshold for nuclear use: "In no way does this approach lower the nuclear threshold. Rather, by convincing adversaries that even limited use of nuclear weapons will be more costly than they can tolerate, it in fact raises that threshold."[21] To some, that rings disingenuous because the NPR insists that Russia is lowering the threshold when it increases the number and prominence of low-yield nuclear weapons. Moreover, both the former commander of STRATCOM and vice-chairman of the Joint Chiefs of Staff, as well as the Air Force Chief of Staff, have acknowledged that lower-yield weapons are more useable and that this can influence the decision-making process for when a president is considering whether to authorize the use of nuclear weapons.[22]

Air-delivered nuclear weapons

The first phase of the nuclear modernization program is focused on air-delivered nuclear weapons. This began with the decision in 2010 to build the B61-12 gravity bomb to replace existing types. Unlike current bombs, the B61-12 is equipped with a guided tail kit to increase the accuracy,[23] and also appear to have some limited earth-penetration capability that could make use of ground-chock coupling to hold underground targets at risk.[24] The first bomb is scheduled to roll off the production line in March 2020, although it might happen later because of production delays of some non-nuclear components. The Air Force plans to produce about 480 B61-12s, which will be compatible with the B-2 and B-21 bombers as well as F-15E, F-16, F-35A, and PA-200 fighter-bombers.

The Air Force currently operates a fleet of 169 heavy bombers: 62 B-1Bs, 20 B-2As, and 87 B-52Hs. Of these, 66 (20 B-2As and 46 B-52Hs) are nuclear-capable, although only 60 (18 B-2As and 42 B-52Hs) are thought to be assigned nuclear delivery roles. It is estimated that there are nearly 850 ALCMs and gravity bombs assigned to strategic bombers, of which about 300 weapons are deployed at bomber bases.

The B-52H is scheduled to operate for several decades more but undergo significant upgrades to electronics, engines, and weapon systems. Development of the next-generation long-range strike bomber, known as the B-21 Raider, is well underway. The stealth-bomber, which together the with B-52H will constitute the future bomber force, looks very similar to the B-2, is expected to make its first flight in 2021 and begin replacing B-1 bombers at Dyess Air Force Base in Texas and Ellsworth Air Force Base in South Dakota from the mid/late-2020s and later B-2 bombers at Whiteman Air Force Base.[25]

To arm its future bomber fleet, the Air Force is developing a new nuclear ALCM known as the LRSO for deployment from 2030. The warhead for the LRSO is known as W80-4, a modified version of the W80-1 that currently arms the ALCM, and is already in development. U.S. defense officials argue that the LRSO is needed to enable bombers to strike targets when faced with advanced air-defense systems and provide flexible strike options in regional scenarios. That might be relevant for the non-stealthy B-52, but the Air Force also says that "the B-21 will be able to penetrate modern air defenses . . . to strike any target worldwide."[26] Some officials say the B-21 needs the LRSO to be able to strike with many missiles at the same time, but many of those targets can probably be held at risk with non-nuclear long-range cruise missiles, such as the extended-range version of the Joint Air-to-Surface Standoff Missile (JASSM-ER), that that the Air Force is buying in large numbers.[27] The U.S. Air Force plans to acquire 1,000 LRSO missiles, of which about half will be nuclear-armed and the rest used for test launches.[28]

Land-based ballistic missiles

The Air Force also operates a force of 400 deployed Minuteman III intercontinental ballistic missiles (ICBMs). But it has 450 silos across three missile wings, with the extra 50 empty silos kept "warm" and can be reloaded with stored missiles if necessary.[29] Each ICBM is loaded with one warhead: either a 335-kt W78/Mk12A or a 300-kt W87/Mk21. Missiles carrying the W78 can be uploaded with up to two more warheads for a maximum of three multiple independently targetable re-entry vehicles (MIRVs). It is estimated that half of the missiles carry the W78 and half the W87, with another 400 W78 warheads in storage for potential MIRVing for a total of 800 warheads assigned to the ICBM force.

The entire Minuteman III force completed a decade-long upgrade in 2015 that extended its service-life until 2030. Moreover, an upgrade is underway of the W87/Mk21 re-entry vehicle to equip it with a new effective fuze (arming, fuzing, and firing unit).[30]

Starting in 2030, the Air Force plans to begin to replace the Minuteman III with a next-generation ICBM known as the Ground-Based Strategic Deterrent (GBSD). For now, the plan is to continue to deploy 400 ICBMs, so the Air Force plans to buy 642 GBSDs, of which 50 are stored and the rest used for test launches and as spares.[31] The expected cost of developing and producing the GBSD is increasing, and in 2017 it was projected to be around $100 billion, up from an initial projection of $62.3 billion in 2015.[32]

To arm the GBSD, the Air Force previously planned to life-extend the W78 as part of a controversial Interoperable Warhead-I (IW-I) program that would have combined ICBM and Navy warheads. In August 2018, however, the U.S. Nuclear Weapons Council canceled the IW-I program and renamed it W87-1. This reflects that it will use a W87-like plutonium pit with insensitive high explosives instead of the conventional high explosives used in the W78. Potential Navy use is still a hypothetical option but not a priority. The projected cost of the W87-1 program is between $12.8 billion and $16 billion.[33]

Although the Air Force awarded project development contracts to both Boeing and Northrup Grumman, Boeing in July 2019 announced that it would not be competing for the final contract. The company said the Air Force's competition was not fair.[34] Unless the Air Force modifies the competition to meet Boeing's concerns, the $80-plus billion program could potentially be awarded to Northrup Grumman without competition, which likely would drive up costs further.

The Air Force has rejected life-extending the Minuteman III a second time, even though that is potentially possible. But the Air Force claims the "Minuteman III will have a difficult time surviving in the active anti-access, area denial environment of the future,"[35] presumably a reference to anti-ballistic missile defense capabilities being developed by Russia. If so, it is unclear why that would not also create problems for the Navy's Trident missile, which is scheduled to operate in the same time period.

Ballistic missile submarines

The sea-based leg of the triad of strategic nuclear forces is also undergoing significant modernization. Of the current fleet of 14 Ohio-class ballistic missile submarines (SSBNs), 8 are based at Naval Submarine Base Kitsap in Washington State and 6 at Naval Submarine Base Kings Bay in Georgia. Of the 14 SSBNs, 12 are normally considered to be operational, with the remaining 2 in refueling overhaul at any given time. Around 8 to 10 SSBNs are normally at sea, of which 4 or 5 are on alert in their designated patrol areas and ready to fire their missiles within 15 minutes of receiving the launch order.

Between 2015 and 2017, the navy replaced Trident II D5 SLBMs with an enhanced version known as the D5LE (LE for 'life extended'). The D5LE is equipped with the new Mk-6 guidance system, designed to improve the SLBM's effectiveness, and also carries the improved 90-kt W76–1 warhead with an enhanced targeting capability, or the 455-kt W88 warhead. It is estimated that some 1,920 warheads are assigned to the SSBN fleet, of which about half are deployed on the missiles. Each SLBM can carry up to eight warheads but normally carries an average of four to five warheads, depending on mission requirements. The New START data indicates that the SSBN fleet carried a total of 918 warheads as of March 2019.[36]

As mentioned above, the Trump administration's NPR recommended development of the low-yield W76–2 warhead. The first W76–2 was completed on February 22, 2019, and a small number (possibly 25–50) will be delivered to the navy by September 2019.[37] Although U.S. war plans are long thought to have included limited strike options with a single or small number of SLBMs, the addition of a low-yield W76–2 to the arsenal to deter Russian use of low-yield tactical nuclear weapons (see above description of the NPR) indicates a new way of using strategic missiles in a tactical fashion.

Development of the next-generation SSBN is well underway, both of hull sections and the nuclear reactor. The first boat, to be known as the U.S.S. Columbia (SSBN-826), will be procured in 2021 and is scheduled to sail on its first deterrent patrol in 2031. A total of 12 Columbia-class SSBNs are planned through the mid-2040s, each equipped with 16 ballistic missiles. The new SSBNs will initially carry the Trident II D5LE SLBM, but it will eventually be replaced with a new SLBM. The NPR states the navy "will begin studies in 2020 to define a cost-effective, credible, and effective SLBM that. . . [can be deployed] throughout the service life of the Columbia SSBN."[38] The next-generation SLBM is currently known as SWS (Strategic Weapon System) 534 or DFLE2,[39] which suggests it might be a second life-extension of the Trident II D5.

Non-strategic nuclear weapons

The Air Force has one basic type of non-strategic nuclear weapons in its stockpile, the B61 gravity bomb, which exists in two modifications: the B61-3 and B61-4. A third modification (B61-10) was retired in September 2016.[40] There are an estimated 230 tactical B61 bombs left in the stockpile.

Approximately 150 of the B61 bombs are thought to be deployed at six air bases in five European NATO countries: Aviano AB and Ghedi AB in Italy; Büchel AB in Germany; Incirlik AB in Turkey; Kleine Brogel AB in Belgium; and Volkel AB in the Netherlands. The bombs are in custody of U.S. Air Force munition units, but in wartime could be handed over to the host nation for employment with their national aircraft. This so-called nuclear sharing arrangement supplies the B61 bombs for the F-16 fighter-bombers of the Belgian and Dutch air forces and PA-200 Tornado fighter-bombers used by the German and Italian air forces. Turkish F-16s are also thought to be nuclear-capable, but it is uncertain if the Turkish Air Force still serves an active nuclear strike mission given the political developments in Turkey in recent years. Rumors that the bombs were quietly withdrawn from Incirlik in 2017[41] have not been confirmed. The base still appears to have an active nuclear storage area, Incirlik completed a major security upgrade in 2015, and is included in another upgrade to be completed in 2019.

The remaining 80 B61 bombs (of the 230 total) are stored in the (continental) U.S. for potential use by U.S. fighter-bombers in support of allies outside Europe, including in East Asia.

As mentioned above, the United States is modernizing its gravity bombs inventory through the B61-12 program. NATO has already agreed to the modernization and deployment of the new weapon, beginning in 2022–24.[42] The B61-12 will use the nuclear explosive package of the B61-4, which has a maximum yield of approximately 50-kt, but will be equipped with a new tail kit to increase its accuracy and standoff capability. The B61-12 will be able to destroy hardened targets that can not be destroyed by the B61-3 or B61-4. It will also enable strike planners to select lower yields for existing targets, which would reduce collateral damage.[43]

Integration flight tests of the B61-12 bombs have been taking place for several years on F-15E, F-16 and Tornado fighter-bombers (in addition to the B-2 strategic bomber). The B61-12 will also be integrated on the new F-35A Lightning stealthy fighter-bomber, which is expected to be nuclear certified in around 2024. Belgium, Italy, the Netherlands, and Turkey have decided to acquire F-35A fighter-bombers, although delivery to Turkey has been suspended in reaction to the country's purchase of the Russian S-400 air-defense system. Germany has decided not to acquire the F-35A, but is considering whether to convert the Eurofighter or buy the F/A-18 to replace the PA-200 Tornado in the nuclear role it reaches the end of its service-life in the mid- to late 2020s.

As described in the NPR section above, the Trump administration NPR plans to acquire several new low-yield nuclear weapons that appear to fall into the tactical nuclear weapons category. The W76–2 warhead for the Trident II SLBM – otherwise considered a strategic weapon – is in production and could potentially be deployed as early as 2020 if Congress agrees. Development and deployment of a nuclear sea-launched cruise missile (SLCM) would take much longer, probably most of a decade, if approved by Congress. If so, it would probably be deployed on Virginia-class attack submarines.

Nuclear weapons production complex

The production complex that maintains and produces nuclear weapons and their delivery vehicles is also undergoing significant modernization. Most of the debate currently is focused on the future of the so-class pit-production capacity, that is, the factories that are built to produce new plutonium cores for the nuclear warheads.

Warhead age is a prominent theme in discussions about the future reliability of U.S. nuclear warheads and the industrial capabilities needed to ensure the weapons will work as designed. As part of that debate, various officials have over the years pointed to the risk of plutonium pits developing defects when they age beyond their intended arsenal design life that could undermine confidence in the arsenal. As a result, proposals for modified warheads and increased pit production capacities were frequent. But in 2006, the JASON scientific advisory panel concluded that

> there is no degradation in performance of primaries of stockpile systems due to plutonium aging that would be cause for near-term concern regarding their safety and reliability. Most primary types have erodible minimum lifetimes in excess of 100 years as regards aging of plutonium.[44]

Now, the pit production requirement is increasing once again. The Trump NPR claims the United States needs a capacity to produce at "no fewer than 80 pits per year by 2030."[45] It is uncertain exactly what drives this increased requirement, but part of it appears to be the plan to build significantly modified warheads. They used to be called Interoperable Warheads and be part of the so-called 3+2 warhead plan, according to which the United States would reduce the number of warhead types from eight to five. To do that, NNSA and the laboratories wanted to build larger warheads with better performance margins that could be modified for use on either an ICBM or a SLBM. The new warheads would have to use insensitive high explosives, even though most ballistic missile warheads use conventional high explosives.

The Navy and Air Force were not particularly interested, however, so recently the NNSA and nuclear laboratories have adjusted their plan. What used to be the first Interoperable Warhead (IW1) mixing W78 and W88, is now known as the W87-1 and will be used only on the ICBMs (for now).[46] Since part of the ICBM force is already armed with W87, replacing the remaining W78s will require production of more W87 pits. But the current pit production capacity at Los Alamos is insufficient so NNSA wants to convert the canceled MOX Facility at the Savannah River site into a second pit production facility. This two-prong plan would create a 30-pit capacity at Los Alamos by 2026 and 50-pit capacity at Savannah River by 2030. The feasibility of the plan's cost estimate and timeline was challenged by the study conducted by the Institute for Defense Analysis.[47]

After the W87-1 warhead gets underway, NNSA plans to continue with what it calls the Next Navy Warhead (formerly referred to as the Interoperable Warhead 2 [IW2] or BM-Y) and the Future Strategic Missile Warhead (formerly referred to as IW3 or BM-Z). These warhead concepts are being developed "to support threats anticipated in the 2030s" and would start delivering in 2034 and 2037, respectively, according to NNSA. The Next Navy Warhead is currently estimated to cost $14 billion to $17.4 billion, and the Future Strategic Missile Warhead $14.9 billion to $18.6 billion.[48]

The warhead production complex is currently operating at its highest capacity since the end of the Cold War and is scheduled to get busier in the years ahead. Although some modernizations were designed to reduce the size of the stockpile, such as the B61-12 program, future production of the nuclear SLCM would likely result in an increase for the first time in the post-Cold War era.

Notes

1 Hans M. Kristensen and Matt Korda, "United States Nuclear Forces, 2019: FAS Nuclear Notebook," *Bulletin of the Atomic Scientists*, May 2019, www.tandfonline.com/doi/pdf/10.1080/00963402.2019.1606503?needAccess=true.

2 U.S. Congressional Budget Office, *Projected Costs of U.S. Nuclear Forces, 2019 to 2028*, January 2019, 1, www.cbo.gov/system/files?file=2019-01/54914-Nuclear-Forces.pdf.

3 U.S. Department of Defense, Office of the Secretary of Defense, "Nuclear Posture Review 2018," February 2018, XI, 51–52, https://media.defense.gov/2018/Feb/02/2001872886/-1/-1/1/2018-NUCLEAR-POSTURE-REVIEW-FINAL-REPORT.PDF.

4 Ibid., 21, 38.

5 Ibid., 21.

6 Ibid., 38.
7 Ibid., 54.
8 Ibid., 55.
9 U.S. Department of Energy, National Nuclear Security Administration, "FY2020 Stockpile Stewardship and Management Plan," July 2019, 1–5, www.energy.gov/sites/prod/files/2019/07/f65/FY20SSMP.pdf.
10 U.S. Department of Defense, Office of the Secretary of Defense, "Nuclear Posture Review 2018," 55.
11 Hans M. Kristensen, "W80-1 Warhead Selected for New Nuclear Cruise Missile," *FAS Strategic Security Blog*, Federation of American Scientists, October 10, 2014, https://fas.org/blogs/security/2014/10/w80-1_lrso/.
12 U.S. Department of Defense, Office of the Secretary of Defense, "Nuclear Posture Review 2018," 55.
13 Ibid.
14 Ibid.
15 Ibid.
16 Hans M. Kristensen, "The Flawed Push for New Nuclear Weapons Capabilities," *FAS Strategic Security Blog*, June 29, 2017, https://fas.org/blogs/security/2017/06/new-nukes/.
17 Hans M. Kristensen, "Increasing Nuclear Bombers Operations," *FAS Strategic Security Blog*, September 25, 2016, https://fas.org/blogs/security/2016/09/nuclear-bomber-operations/.
18 For a history of the international disputes over nuclear-capable ship visits during the Cold War, see: Hans M. Kristensen, "The Neither Confirm Nor Deny Policy: Nuclear Diplomacy at Work," *Federation of American Scientists*, February 2006, www.nukestrat.com/pubs/NCND.pdf.
19 Seth Robson, "U.S. Navy to Return to New Zealand After 30-Year Rift Over Nukes," *Stars and Stripes*, July 21, 2016, www.stripes.com/news/us-navy-to-return-to-new-zealand-after-30-year-rift-over-nukes-1.420187.
20 Hans M. Kristensen, "Russian Nuclear Forces: Buildup or Modernization?" *Russia Matters*, September 14, 2017, www.russiamatters.org/analysis/russian-nuclear-forces-buildup-or-modernization.
21 U.S. Department of Defense, Office of the Secretary of Defense, "Nuclear Posture Review 2018," II.
22 Hans M. Kristensen, "General Cartwright Confirms B61-12 'Could Be More Useable'," *FAS Strategic Security Blog*, November 5, 2015, https://fas.org/blogs/security/2015/11/b61-12_cartwright/; Hans M. Kristensen, "General Confirms Enhanced Targeting Capabilities of B61-12 Nuclear Bomb," *FAS Strategic Security Blog*, January 23, 2014, https://fas.org/blogs/security/2014/01/b61capability/.
23 Hans M. Kristensen, "Capabilities of B61-12 Nuclear Bomb Increase Further," *FAS Strategic Security Blog*, October 30, 2013, https://fas.org/blogs/security/2013/10/b61-12hearing/; Hans M. Kristensen, "B61 LEP: Increasing NATO Nuclear Capability and Precision Low-Yield Strikes," *FAS Strategic Security Blog*, June 15, 2011, https://fas.org/blogs/security/2011/06/b61-12/.
24 Hans M. Kristensen, "Video Shows Earth-Penetrating Capability of B61-12 Nuclear Bomb," *FAS Strategic Security Blog*, January 14, 2016, https://fas.org/blogs/security/2016/01/b61-12_earth-penetration/.
25 U.S. Air Force, Secretary of the Air Force Public Affairs, "Air Force Selects Locations for B-21 Aircraft," May 2, 2018, www.af.mil/News/Article-Display/Article/1510408/air-force-selects-locations-for-b-21-aircraft/.
26 U.S. Air Force, "United States Air Force Acquisition Report Fiscal Year 2018: Cost-Effective Modernization," 2019, 23, www.af.mil/Portals/1/documents/5/FY18_AQReport.pdf.

27 For background and context on the LRSO see: Hans M. Kristensen, "LRSO: The Nuclear Cruise Missile Mission," *FAS Strategic Security Blog*, October 20, 2015, https://fas.org/blogs/security/2015/10/lrso-mission/; Hans M. Kristensen, "Forget LRSO: JASSM-ER Can Do the Job," *FAS Strategic Security Blog*, December 16, 2015, https://fas.org/blogs/security/2015/12/lrso-jassm/; Kingston Reif, "Examining the Flawed Rationale for a New Nuclear Air-Launched Cruise Missile," *Arms Control Today, Issue Briefs* 8, no. 2 (June 12, 2016), www.armscontrol.org/Issue-Briefs/2016-06-12/Examining-the-Flawed-Rationale-for-a-New-Nuclear-Air-Launched-Cruise-Missile.

28 General Robin Rand, Commander Air Force Global Strike Command, "Prepared Statement to the U.S. Senate Armed Services Committee, Strategic Forces Subcommittee on the FY19 Posture for Department of Defense Nuclear Forces," April 11, 2018, 13, www.armed-services.senate.gov/imo/media/doc/Rand_04-11-18.pdf.

29 "USAF Removes Last of 50 Minuteman III ICBMs and Meets NST Requirements," *Air Force Technology*, July 3, 2017, www.airforce-technology.com/news/newsusaf-removes-last-of-50-minuteman-iii-icbms-to-meet-nst-requirements-5860111/.

30 Michael Padilla, "Sandia on Target for First Mk21 Fuze Flight Test in 2018," *Sandia Lab News* 70, no. 6 (March 16, 2018), www.sandia.gov/news/publications/labnews/_assets/documents/issues/2018/labnews03-16-18.pdf.

31 Kingston Reif, "Air Force Drafts Plan for Follow-on ICBM," *Arms Control Today*, July 8, 2015, www.armscontrol.org/ACT/2015_0708/News/Air-Force-Drafts-Plan-for-Follow-on-ICBM.

32 Kingston Reif, "New ICBM Replacement Cost Revealed," *Arms Control Today*, March 2017, www.armscontrol.org/act/2017-03/news/new-icbm-replacement-cost-revealed.

33 U.S. Department of Energy, National Nuclear Security Administration, "FY2020 Stockpile Stewardship and Management Program," 8–41.

34 Rachel S. Cohen, "Boeing Backs Out of Nuclear Missile Competition, Prompting USAF Choices," *Air Force Magazine*, July 25, 2019, www.airforcemag.com/Features/Pages/2019/July%202019/Boeing-Pulls-Out-of-Ground-Based-Strategic-Deterrent-Program.aspx.

35 U.S. Air Force, Air Force Nuclear Weapons Center Public Affairs, "AF Releases New ICBM Solicitation," July 29, 2016, www.af.mil/News/Article-Display/Article/881075/af-releases-new-icbm-solicitation/.

36 U.S. Department of State, "New START Treaty Aggregate Numbers of Strategic Offensive Arms," *Fact Sheet*, July 1, 2019, www.state.gov/new-start-treaty-aggregate-numbers-of-strategic-offensive-arms-10/.

37 U.S. Department of Energy, National Nuclear Security Administration, "NNSA Completes First Production Unit of Modified Warhead," February 22, 2019, www.energy.gov/nnsa/articles/nnsa-completes-first-production-unit-modified-warhead; Dan Leone, "NNSA Has Started Building Low-Yield Sub Warhead," *Nuclear Security & Deterrence Monitor* 23, no. 4 (January 25, 2019), www.exchangemonitor.com/nnsa-started-building-low-yield-sub-warhead/.

38 U.S. Department of Defense, Office of the Secretary of Defense, "Nuclear Posture Review 2018," 49.

39 U.S. Department of Defense, "U.S. Nuclear Deterrence Policy," *Fact Sheet*, April 1, 3, https://media.defense.gov/2019/Apr/01/2002108036/-1/-1/1/U.S.-NUCLEAR-WEAPONS-CLAIMS-AND-RESPONSES.PDF; James Peterson, "Navy Strategic Missile Boss Starting Concept Development for New Missile," *Seapower*, May 24, 2017, http://seapowermagazine.org/stories/20170524-Benedict.html.

40 U.S. Department of Energy, National Nuclear Security Administration, "Fiscal Year 2018 Stockpile Stewardship Management Plan," November 2017, 1–13, www.energy.gov/sites/prod/files/2017/11/f46/fy18ssmp_final_november_2017%5B1%5D_0.pdf.

41 Joseph Hammond, "The Future of Incirlik Air Base," *Real Clear Defense*, November 30, 2017, www.realcleardefense.com/articles/2017/11/30/the_future_of_incirlik_air_base.html.
42 For an overview of the modernization of U.S. and NATO non-strategic nuclear forces, see: Steve Andreasen et al., *Building a Safe, Secure, and Credible NATO Nuclear Posture*, Nuclear Threat Initiative, January 2018, www.nti.org/media/documents/NTI_NATO_RPT_Web.pdf.
43 Kristensen, "Video Shows Earth-Penetrating Capability of B61-12 Nuclear Bomb."
44 The MITRE Corporation, *Pit Lifetime*, JSR-06-335, November 20, 2006, 1, https://fas.org/irp/agency/dod/jason/pit.pdf.
45 U.S. Department of Defense, Office of the Secretary of Defense, "Nuclear Posture Review 2018," 64.
46 U.S. Department of Energy, National Nuclear Security Administration, "W78 Replacement Program (W87-1): Cost Estimates and Use of Insensitive High Explosives," December 2018, https://nukewatch.org/newsite/wp-content/uploads/2019/02/W78-Replacement-Program-Cost-Estimates-IHE-1.pdf.
47 "New Study Finds NNSA Pit Strategy Unlikely to Meet Production Deadline, Lawmaker Says," *Exchange Monitor*, May 2, 2019, www.exchangemonitor.com/235121-2/.
48 U.S. Department of Energy, National Nuclear Security Administration, "FY2020 Stockpile Stewardship and Management Program," 2–37, 2–45, 8–6, 8–41.

3 Russia's nuclear modernization

Philip M. Baxter

The Russian Federation is in the midst of a multi-decade-long modernization of its strategic and non-strategic nuclear forces. With decisions to proceed down this route beginning in the 1990s, the modernization efforts of the Russian nuclear arsenal have continued at pace since the signing of New START in 2011. In addition to updating Soviet-era nuclear systems, Russia has sought to advance new nuclear warheads and launchers, modernizing and upgrading every leg of the Russian nuclear triad to include strategic bombers, sea-based missiles, and land-based missiles.

Since the collapse of the Soviet Union, nuclear weapons have underwritten Russian security and arguments for its great power status. During this period, Russia's conventional military power has been somewhat of a laggard, and even today remains deficient. As such, Russian strategy and doctrine have adapted to these realities, as the recent U.S. *Nuclear Posture Review* points out, emphasizing the potential coercive uses of nuclear weapons.[1] The policy of 'escalating to de-escalate', or escalating to win, depending on the interpretation, is coupled on the conventional side by reliance on asymmetric capabilities to compensate for conventional power shortfalls. This asymmetric focus has been on display in Russia's push for greater regional influence over the last decade in Estonia, Georgia, and Ukraine. That said, Russia has also sought to upgrade its conventional forces as well.

While developments in Russian military strategy and doctrine, as well as conventional forces, are of critical importance to understanding the outlook for U.S.-Russia strategic relations, this chapter will focus primarily on reviewing efforts made in the nuclear weapons domain. This will include a discussion of upgrades to existing systems and developments of new weapon systems. The chapter will proceed in three parts. First, we will review the modernization efforts directed at improving legacy systems from the Soviet era. Next, we will examine new developments, including the development of new intercontinental range systems, hypersonic weapons, a new nuclear torpedo, a new ground-launched cruise missile, and upgrades to missile defense systems. The final section of this chapter will, within the context of these modernization efforts, review some of the implications of these developments.

Legacy systems

This section will review strategic systems developed during the Soviet Union era which remain in service as of 2018. The focus here is to establish the context and track the evolution of Russia's nuclear arsenal capacity from the Soviet era to the present, examining intercontinental ballistic missiles, submarines and sea-launched missiles, and strategic bombers.

Intercontinental ballistic missiles

At the peak of the Cold War, the Soviet Union had stockpiled reportedly 45,000 nuclear warheads and had deployed a variety of missile systems during the nearly 50 years of vertical proliferation between the United States and Russia.[2] In terms of intercontinental delivery systems, the Soviet Union developed and deployed roughly 16 classes of missiles, with five missile categories active at the dissolution of the Soviet system: the NATO-designated SS-18 (or Soviet-designated R-36M or treaty-designated RS-20),[3] the SS-25 (or RT-2PM or RS-12M Topol), the SS-24 (or RT-23 or RS-22), the SS-17 (or UR-100MR or MR-UR-100 Sotka), and the SS-19 (or UR-100N or RS-18A). The number of deployed missile systems further contracted during the early post-Soviet period due to the signing of the Strategic Arms Reduction Treaty in 1991, with its entering into force in 1994. The SS-17 would be phased out in the early 1990s and decommissioned in 1993. The silo- and rail-based SS-24 would halt production after the split-up of the Soviet Union – as its production facilities were located in Ukraine – and would be decommissioned through the 2000s, with its elimination completed by 2008. The remaining legacy systems, the SS-18, the SS-25, and the SS-19, would all continue to serve as the foundation of Russia's strategic deterrent through the 2000s and 2010s.

While this analysis will review each of the different systems, one thing of note that is historically important is that much of the capacity for missile construction was housed in Ukraine, while maintenance servicing support was maintained in Russia. As such, much of the capacity for missile development was lost by Russia after the Soviet Union collapsed.

The SS-19, also known as the UR-100N within the Soviet Union, RS-18 bilaterally, or Stiletto by NATO, was a key component of the Soviet nuclear arsenal. The two-stage, liquid-propellant, silo-based missile entered into service in 1980. Similar to the previously deployed SS-11, the SS-19's increased size enabled a throw-weight, or payload capability, of three times that of its predecessor.[4] Its larger size allowed the missile to handle up to six warheads, although START I required all of these weapons to be reconfigured to carry only one warhead.[5] This, in addition to START II, which moved the maximum operation inventory of the SS-19 from 360 missiles in 1984 to 105 permissible after the treaty, further reduced the number in the arsenal.[6] The missile, at 27 meters in length and 2.5m in diameter, has a range of roughly 10,000km and a circular error probable (CEP) measure, or a median error radius, of 550m.[7] The missile is being replaced by the also silo-based SS-27 Mod 2 (or RS-24) and is scheduled to be retired, and as such only roughly 20 remained deployed.[8] That said, the SS-19 was used as the launch vehicle in the recent test of the new aero-ballistic hypersonic Avangard missile in October 2017, and while the RS-28 Sarmat will replace the SS-19 and will likely be the carrier of the hypersonic missile in the future, this may extend the SS-19 past initial estimates.[9]

Another silo-based missile, the SS-18 was initially developed in the late 1960s and early 1970s, and went through six variations during its initial deployment in 1975 through to the present. The two-stage, liquid-propellant inertial guided missile was designed as the replacement to the SS-9 ICBM, which was in service between 1967 and 1978, and would be a "bulwark of the [Strategic Rocket Force's] hard-target-kill capability."[10] The missile ranged from 33.6m to 37.25m, depending on the variant, with a diameter of 3m, and a payload between 7.2t and 8.8t.[11] The original SS-18 (Mod 1) was a single warhead weapon with a yield between 18 and 25 megatons. Mods 2 and 3 would be a MIRVed version of the initial missile with an ability to carry up to eight warheads, and an updated single-warhead missile with a greater range, from 11,200kms to 16,000kms, and accuracy.[12] Mod 4 would be an improvement on Mod 2, with a larger range and greater accuracy of the multiple independently targetable re-entry vehicle (MIRVed) weapon. The more recent upgrades to the SS-18 were Mods 5 and 6, referred to as the R-36M2 Voivode or RS-20V. With an initial operational capability in 1988 and 1990, respectively, these missiles remained in service as of 2018. These latest variants provided greater accuracy and warhead yield capacity over previous versions of the missile. That said, these missiles are being gradually retired, with approximately 46 SS-18s with 460 warheads remaining.[13] Like the SS-19, the SS-18 will be replaced by the RS-28 Sarmat (or SS-x-29) missile.

Finally, the SS-25, or known in bilateral agreements as the RS-12M or RT-2PM Topol within Russia, is a road-mobile, three-stage, solid-fuel missile. Developed through the mid-1970s and mid-1980s, the missile was deployed in 1988. The single-warhead missile was roughly 21.5m in length, 1.8m in diameter, and had a yield of 550-kt.[14] The breakup of the Soviet Union created significant problems for the SS-25, as the transporter erector launcher (TEL) for the missile was manufactured in Belarus and much of the guidance system was manufactured in Ukraine.[15] Additionally, as road-mobile missiles have significantly higher maintenance and personnel costs associated with them than that of siloed weapons, the deployment of these missiles was scaled back as resources became more limited. This missile is being retired and replaced by the SS-27 Mod 1 and 2. Kristensen and Norris observed that there is uncertainty in terms of how many SS-25 remain operational, as the upgrades to the arsenal consist of simply switching out the missile under camouflage nets, rather than massive upgrades to silos or other infrastructure, which limits the observable changes via satellite imagery.[16] They estimate that roughly 72 SS-25 remain deployed.[17]

Submarine-launched ballistic missiles

Russia still maintains one legacy SLBM, the SS-N-18 (R-29R, R-2S, or RSM-50) and the SS-N-23 (R-29RM or RSM 54), both based on the SS-N-8 (R-29 or

RSM-40) missile. Like ICBMs, while Russia has continued to maintain several missiles through lifecycle extension programs, they have focused primarily on the development of new strategic weapons. With regards to SLBMs, this is likely due to three factors: cost, loss of infrastructure to other states after the collapse of the Soviet Union, and the introduction of a new class of submarine, the Borei-class, which has different specifications than the previous versions.

The SS-N-18 was the Soviet Union's first SLBM with an intercontinental range up to 8,000km and was deployed on Kalmar class (Delta III) submarines. Initially, a single warhead weapon, variations of the weapon were designed to handle up to seven warheads, although the seven-warhead version was never deployed – a three warhead MIRV system was.[18] Deployed in 1979, the two-stage, liquid-propellant missile would deploy on fourteen Kalmar/Delta III submarines.[19] This early variant of the missile carried a single 450-kt warhead, while the subsequent Mod 2 was equipped with three 200-kt warheads.[20] As of 2018, three Delta III submarines carrying SS-N-18 remained in service but with only two deployed at any time.[21] The SS-N-18 will be replaced by the SS-N-32 Bulava missile for the Borei-class submarines.[22]

The SS-N-23 (R-29RM or RSM 54) is deployed on Delta IV submarines, entering operational service in 1986. The missile has a payload capacity of four 100-kt MIRV warheads, with a range of 8,300 km.[23] It is estimated that 96 SS-N-23 missiles are in the Russian arsenal.[24] The SS-N-23 underwent a lifecycle extension program beginning in 1998 to further push out the operational life of the missile and its deployment on the Delta IV submarines. As the Delta IVs will be replaced by the Borei class submarine, which will utilize 16 SS-N-32, it is likely that this missile system will be phased out eventually.

Russia operates two legacy submarine classes, which makes up eight of its current eleven operational nuclear-powered ballistic missile submarines. The Delta III (667BDRM) and Delta IV (667BDRM) were first entered into service in 1983 and 1986, respectively. Each submarine can carry up to 16 submarine-launched ballistic missiles, either the SS-N-18 or SS-N-23, depending on the vessel. As mentioned previously, The Delta III submarines are nearing phase-out by the new Borei class submarine, and the Delta IV's will likely remain in service for several more years until additional Borei class can be brought online.[25]

Heavy bombers

Russia currently operates two heavy bombers which are nuclear-capable from the Soviet era: the Tu-160 Blackjack and the Tu-95MS Bear. The Tu-95MS, based on the Tu-142 design, has an operational range of 10,500km and could carry up to 16 of the Kh-55 (AS-15A) air-launched cruise missiles, each with a yield of 200-kt, or gravity bombs.[26] The Tu-160 Blackjack is a multi-mission strategic bomber capable of handling a variety of munitions, including cruise missiles, short-range guided missiles, and gravity bombs, both nuclear and conventional. It can carry 12 Kh-55 (AS-15B), either 12 or 24 Kh-15(AS-16) hypersonic aero-ballistic missiles, or gravity bombs.[27] The Tu-95MS Bear entered into service in 1981, while the TU-160 Blackjack began in 1987. Both heavy bombers are going through a series of modernization and upgrades, including making them compatible with the Kh-101 (AS-23A), a conventional cruise missile, and the Kh-102 (AS-23B), a nuclear cruise missile.[28] This modernization program will enable the Tu-95MS Bear and Tu-160 Blackjack to continue operating beyond 2030.[29] At present, somewhere in the range of 60–70 bombers of each type are currently in the Russian inventory.[30]

Legacy summary

This section has briefly reviewed strategic systems developed during the Cold War period which still serve to underline Russia's nuclear deterrent. One thing not covered in the above discussion is the Russian nuclear weapon stockpile. Russia continues to retain a sizable nuclear stockpile from the Cold War period, but also has significant weapons infrastructure and production capabilities so as to produce new, large quantities of nuclear weapons each year.[31] Kristensen and Norris estimate that Russia has 1,600 nuclear weapons deployed, with a reserve of 2,750, roughly 2,500 awaiting disarmament (as of 2018), for a total inventory of 6,850.[32]

From the above discussion, several points are evident regarding how Russia has approached its strategic arsenal in the post-Cold War period. First, Russia is determined to maintain, although significantly less diversified in its delivery systems, all components of its nuclear triad: land-based intercontinental ballistic missiles, submarine-launched ballistic missiles, and strategic bombers. The draw-down of the different type of weapons in each of the above components enabled significant cost savings in maintenance and upkeep, while still allowing for an arsenal capable of managing a nuclear deterrence strategy.

A second takeaway is that, even with the more limited scope, the decision in the late 1990s to modernize particular elements of the nuclear arsenal was focused on maintaining numerical parity with the U.S. and their production capabilities.[33] As was mentioned previously, the loss of production capacity after the Soviet Union's collapse put significant strain on the ability of Russia to maintain a diverse deterrent as well as to innovate. As a result, lifecycle management programs were needed to extend the legacy system until a future date. The rebound of the Russian economy and resurgence of Russia is asserting itself within its previous sphere of influence, such as in the case of Estonia in 2007, Georgia in 2008, and Ukraine 2007–2008.

New developments

Since the signing of New START in 2010, Russia has continued to modernize its nuclear forces, rolling out investments made in the 2000s that move beyond the previous lifecycle extension efforts directed at the legacy systems which seek to continue the operability of the systems. Rather, renewed efforts and capital investment directed at new capabilities have been rolled out to replace, or are scheduled to replace, currently deployed forces over the next few years. The level of this investment was clear when in 2011, Vladimir Popovkin, Russia's First Deputy Minister of Defense, announced that Russia would invest nearly U.S. $70 billion into strategic nuclear forces over the next decade.[34] This strategic modernization harks back to the Cold War where improving weapon systems was seen as critical in ensuring nuclear deterrence, or providing an advantage during periods of escalation. As Russia has indicated, these emerging weapons are designed to counter the deployment of a U.S./NATO missile defenses, attempting to make such efforts ineffective and providing Russia with an escalatory gain.[35] As was pointed out in the 2018 *Nuclear Posture Review*, the development and deployment of new strategic systems align with the shift in Russian nuclear policy, which envisions escalating to de-escalate – or escalating to win – depending on your translation.[36] This shift is markedly different from the 20 years or so which followed the end of the Cold War, harking back to the arms race between the 1970s and 1980s where both the United States and the Soviet Union feared for their own existential security, and consequently engaged in what was perceived as security-increasing behavior.[37]

Intercontinental ballistic missiles

After the collapse of the Soviet Union, Russia rolled out one new ICBM, the RT-2PM2 Topol-M, also known as the SS-27 Mod 1 or RS-12. This is a three-stage, solid-propellant, silo-based or road-mobile intercontinental ballistic missile. The Topol-M carries a single warhead with an 800-kt yield but the design is compatible with MIRV warheads. Deployment of the SS-27 Mod 1 began in 1997 and was completed with a total of 78 missiles put into service by 2012.

That said, the current emphasis of the Russian ICBM modernization efforts has been the SS-27 Mod 2, also known as the RS-24 or Yars, and the SS-X-29. The SS-27 is an upgraded SS-27 Mod 1, enabling it to carry up to four MIRVs. Starting its deployment as a mobile ICBM in 2010, with the silo version entering into service in 2014, the missile is currently deployed in five Strategic Rocket Force divisions, with another five currently being upgraded.[38]

Russia has also been developing the SS-X-29, also referred to as the RS-28 Sarmat, which is intended to replace the SS-18 (RS-20), giving it the nickname, "Son of Satan." The new missile is a liquid-fueled, silo-based heavy ICBM. Following repeated setbacks in its development, flight testing began in late 2017, likely pushing back the expected in-service date of 2020.[39] The SS-X-29 was one of six new weapon systems that were unveiled by Russian President Vladimir Putin in March 2018.[40]

The SS-X-29 is notable for several reasons. First, its payload capacity enables it to carry a larger quantity of warheads if restrictions on the number of warheads per missile stemming from current arms control agreement lapse. Second, the

SS-X-29 is slated to carry the new hyper-glide vehicle, another one of the weapons unveiled by President Putin in 2018. The Avangard hypersonic glider can reportedly travel at Mach 27 speeds and can maneuver both the pitch and yaw in order to avoid missile defense systems.[41] The SS-X-29 would reportedly be able to carry up to 24 of these gliders, which can carry either nuclear or conventional warheads.

Two other ICBMs have also been under development, but appear to be suspended as of mid-2018. The BZhRK Barguzin, or SS-X-32Zh, is a rail-mobile ICBM which began development in 2013. It was suspended in late 2017 to focus resources on the SS-X-29.[42] The second system, the RS-26 Rubezh, a small, lightweight road-mobile ICBM, was suspended in March 2018. This system had been tested and was ready for deployment as of March 2015, but had not entered into services, perhaps due to U.S. concerns regarding its potential for modification that would allow it to circumvent the INF Treaty.[43]

Submarines sea-launched ballistic missiles

In the post-Soviet era, Russia has focused its efforts on its submarine fleet. Since 2003, work that had stalled after the Cold War restarted and roughly U.S. $78 billion dollars was invested into the Russian Navy.[44] Given the rapidly aging status of many of the sea-launched ballistic missiles and submarine fleet, significant investment was placed in lifecycle extension of a few systems, while introducing modernized replacements set for a staged rollout throughout the late 2010s and into the 2020s.

Russia has plans to build five additional Borei-class submarines to join the three in service. These five ships will likely be an improved design over the initial Borei class, and construction will be based on a design called the Borei-A or Project 995A.[45] There are also reports that six additional Borei-class SSBNs will be acquired, moving the total SSBN fleet size to 14 if the dated Delta III and Delta IV class vessels are retired.[46]

Each of the Borei-class SSBNs can carry up to 16 SLBMs. Currently deployed Borei-class, in addition to forthcoming SSBN deployments, carry the SS-N-32 or RSM-56 Bulava. The SS-N-32 has faced significant challenges in its development, including a failure rate of over 50% between 2004 and 2009.[47] After modifications and another round of tests, the missile entered into service in 2018.[48] Each SS-N-32 can carry up to 6 warheads with a yield of 100-kt each.

Beyond this, there are additional plans to develop a new class of nuclear submarine to serve roles as both an attack submarine and a cruise missile carrier. The Husky-class nuclear submarine would be a fifth-generation submarine and would, reportedly, be able to carry the 3M22 Zircon, a scramjet-powered maneuvering anti-ship hypersonic cruise missile. The high speed, at potentially Mach 8, would make it significantly faster than prior cruise missiles, making it very difficult to track and intercept.[49] The construction of the first models of this new class would likely come in the 2020s.

Heavy bombers

As discussed previously, both the Tu-160 Blackjacks and Tu-95 serve as the two heavy bombers for nuclear missions. Most of these aircraft are going through upgrades which will extend their lifespan. Additional upgraded Tu-160 Blackjacks, known as the Tu-160M2, will also be constructed in order to fill the void stemming from delays in the development of a new heavy bomber.[50] This bomber, known as the PAK-DA, will likely serve as the ultimate replacement for the Tu-95MS and Tu-160.[51] Research and development of the new bomber began in 2014, with the goal to start tests by 2021, but will likely not occur until 2026, and then deployment three years after.[52] Kristensen and Norris have noted that the Russian aviation industry lacks the "capacity to develop and produce two strategic bombers at the same time."[53]

INF killer

While this chapter has focused primarily on Russia's strategic nuclear arsenal, given the increasing focus on the development of new technologies and modernizing weapon systems, it is worth noting the linkages between Russian nuclear modernization efforts and the impact on weapon treaties between the United States and Russia. In particular, Russian efforts to expand its nuclear delivery capability has been viewed by the U.S. as being in violation of the Intermediate-Range Nuclear Forces Treaty, or INF Treaty. As a result, U.S. Secretary of State Mike Pompeo announced on February 1, 2019, that the U.S. had provided formal notice that it would withdraw from the INF Treaty effective in six months, in accordance with Article 15 of the treaty.[54] The day following the announcement of U.S. departure from the treaty, President Putin stated that Russia was also suspending its obligations under the treaty.[55]

The INF Treaty was a unique development in the history of arms control as it eliminated an entire class of weapon systems: land-based ballistic and cruise missiles with a range between 500–5,500km. Entering into force in 1988, it did not include air- and sea-launched missiles within that range but did result in the United States and Soviet Union/Russia decommissioning 2,692 missiles. It has been noted as being a fundamental pillar of European security for the past 30 years.[56]

While an effective treaty that removed thousands of weapons from a potential conflict, placing greater priority on strategic parity, the INF Treaty has encountered extensive concerns from both parties since the early 2000s. The United States has suspected violations by Russia since the late-2000s/early 2010s, having raised official concerns with the Russian Security Council in 2013, due primarily to the development of the Novator 9M729, or NATO designated SSC-8.[57] The land-based cruise missile is thought to have a range that falls between 500km and 5,500km, a violation of the INF Treaty. Additionally, Russia has long believed the INF Treaty had lost its strategic relevance and proposed joint withdrawal;[58] the argument largely stemming from missile advances by other states in Eurasia, and thus, in the range of Russia.

U.S. Director of National Intelligence Daniel Coats noted in November 2018 that Russia is likely to have started the development of the missile, which can carry either a conventional or nuclear warhead, in the mid-2000s and conducted flight tests through that period to 2015.[59] Researchers have pointed out that the SSC-8 was developed under the guise of a sea-launched cruise missile program since about 2008. Others have observed that the SSC-8 is derivative of the Kalibr sea-launched cruise missile, which is allowed by the INF Treaty. With an identical missile tested from a ground-fixed position, but a dedicated sea-launched missile test facility, it is evident that through much of the development of this missile, which launches from a ground-mobile platform, that Russia was attempting to mask its intentions.

The one major question that has persisted over the last few years, and remains the point of contention, is the actual ground-based range of the missile. As a full-range of tests have not been conducted, we are left to make estimates. Researchers have assessed the missile's range as being inside the range restriction by the INF Treaty, between 500 and 5,500km, and would be in violation of the treaty. While tensions over this issue rapidly advanced in 2018 and early 2019, it is likely that this missile will still require several years before entering service and being deployed.

In addition to announcing its departure from the INF Treaty, the U.S. declared recently that they would also begin to develop the component necessary for an intermediate-range ground-based ballistic missile.[60] These developments coupled with the general outlook on arms control agreements by the current U.S. administration, as well as the need for renegotiation of the New START Treaty, make it increasingly less likely that the INF Treaty will be revived.

Other developments

A March 1, 2018, presentation in Moscow by Russian President Vladimir Putin sought to showcase Russian strength and innovation. The speech resembled the language of the Cold War period, with its boisterous posturing, and overall attempt to convey notions of strength through hard power and technological superiority.[61]

During the event, Putin presented a series of technological advances within its strategic weapons portfolio. As previously discussed, the presentation included the RS-28, or Sarmat, a new intercontinental ballistic missile with a larger payload and capability to carry hypersonic glide vehicles. This type of system was also showcased, with Putin highlighting two variants of hypersonic weapons. The first variant, which was discussed previously, was the hypersonic weapon which is air-launched and can travel at very high speeds while remaining highly maneuverable.[62] The second, which was not visually demonstrated during the presentation, was a space-based projectile. A laser system was also described, which is presumably an anti-aircraft or anti-missile system.[63] Putin stated that deployment of this laser system had begun the previous year. Finally, a simulation of a cruise missile design equipped with a nuclear-propulsion engine, allowing the missile to operate with unlimited range, was shown, as well as the development of an underwater, nuclear-powered drone.

Conclusion: why these developments matter

This chapter has attempted to briefly summarize Russian strategic nuclear forces and developments over the last thirty years. In the immediate period following the Cold War, resource limitations severely impacted the Russian military, leading to a period of rapid constriction in size and stagnation in innovation in military arms. This led to a heavy dependence on its aging nuclear forces to ensure Russia's security. This was obviously recognized as many of the first modernization efforts were directed at its nuclear forces, enabling the deployment of a few upgraded weapon systems in the late 1990s and early 2000s. Many of the efforts to modernize the Russian military extend beyond their strategic arsenals, including reorganizing its forces and major investments into rearming its military.[64]

After the directed effort following the Cold War to extend the lifecycle of existing systems – an effort marked by ensuring state security while scaling back its military forces and restricting resources – two distinct periods in Russia's modernization efforts in their strategic forces can be linked to actions by the United States. The first occurred in the early 2000s. Russia saw the U.S. invasion of Iraq and Afghanistan, a period in which the United States was able to swiftly defeat conventional forces and overthrow the existing governments with very limited causalities. With a defense budget significantly smaller than the United States for the decade preceding this, as well as forces that have been degraded, even conservative strategists would conclude that Russia's military would be unable to compete with the U.S. in the traditional arena.

The second period, occurring in the early 2010s, was driven more by opportunism than strategic need. This was based on the lackluster response from the United States following the invasion of Georgia and asymmetric actions against Estonia, coupled with further reductions in strategic arms stemming from the New START agreement, and U.S. pivot away from Europe to Asia as its primary strategic focus. Such shifts have provided Russia flexibility in adapting its strategy in its near-abroad vicinity that has enabled provocative actions at the non-strategic levels while limiting response due to parity at the strategic level. These provocative actions have included asymmetric approaches to warfare that enable territorial gains and further its desire to project power regionally, while handcuffing policy response options by the impacted states, NATO, and the U.S. The continual focus on modernizing the strategic level triad, introducing new weapon platforms, and seeking to upgrade the military more broadly, raises serious international security questions regarding the potential continued asymmetric power acquisition.

Beyond the renewed focus on strategic standing, there are also significant domestic political reasons for modernization efforts into its strategic arsenal. While much of Putin's March 2018 speech discussed domestic issues, clear signals were made that ensuring security was of utmost importance.[65] This focus on security and, more importantly, the return of Russia as a great power on the international stage, has been a critical component of Putin's popularity throughout this tenure as President. Projecting strength is critical. When his domestic popularity drops, bolder talk and action occurs, typically directed at the U.S. or a country on its periphery. Given this pattern, instances in which global events position Russia or Putin in a negative light or if there is a significant downturn in the Russian economy, may result in a more aggressive foreign policy regarding its neighbors.

These developments in strategic arms modernization will also play a critical role in the future of the Strategic Arms Reduction Treaties, the most recent of which being New START. The first U.S.-Russian nuclear arms treaty to include verification protocols since START I, the treaty limited both parties to 700 deployed ICMBS, SLBMS, or heavy bombers with nuclear weapons, 1,550 deployed nuclear warheads, and 800 deployed and non-deployed ICBM launchers, SLBM launchers, and heavy bombers for nuclear missions. The treaty is set to expire 10 years from entry into force – in February 2021. With the increased focus by Russia in modernizing its nuclear arsenal, in developing new weapon platforms to fulfill missions previously held by strategic nuclear armaments, the linkages between Russian domestic politics and strength projection, and the outlook of the present administration regarding arms control agreements, limited progress is likely until shortly before the treaties expiration. As of late 2019, limited progress has been made in renegotiating treaty renewal, in particular the U.S. has not yet established a negotiating position regarding the treaty, and this is unlikely to advance further in the wake of the INF Treaty collapse.[66] The potential collapse of bilateral nuclear treaties between the U.S. and Russia, along with the introduction of new weapon systems through modernization efforts, could result in significant difficulties in the strategic stability model between the two countries.

In sum, this chapter has sought to review the evolution of Russia's nuclear arsenal over the last two decades, with a particular emphasis on the future of its strategic deterrent. Through the course of this discussion, a particularly troublesome theme that has persisted throughout is the continued, as well as increasing value to which strategic nuclear arms play in Russia. During periods of austerity, but also peace, a focus was placed on ensuring resources were available for improvements and lifecycle extension for the core of its nuclear weapons program. Having extended these platforms from the Soviet era, as well as making investments in the 1990s to avoid gaps in service, the further modernization and advancement of the legs of the triad are likely to stall further arms agreements with the United States for further reductions. The demise of the INF Treaty and the rapidly approaching end date for the New START Treaty, coupled with further technological investments and the U.S. slowly ramping up its modernization efforts, cast a dark shadow upon the future of potential arms racing between these two seemingly renewed adversaries.

Notes

1 U.S. Department of Defense, Office of the Secretary of Defense, "Nuclear Posture Review 2018," February 2018, https://media.defense.gov/2018/feb/02/2001872886/-1/-1/1/2018-nuclear-posture-review-final-report.pdf.

2 Natural Resources Defense Council, "Global Nuclear Stockpiles, 1945–2006," *Bulletin of the Atomic Scientists* 62, no. 4 (2006), 65.

3 Throughout this chapter, weapon systems will follow this naming convention: identifying their NATO-designation first, then their Soviet-designation, and finally any treaty designations for the weapon system.
4 John Pike et al., "WMD Around the World: UR-100N/SS-19 STILLETO," *Federation of Atomic Scientists*, last modified July 29, 2000, https://fas.org/nuke/guide/russia/icbm/ur-100n.htm.
5 Treaty Between the United States of America and the Union of Soviet Socialist Republics on the Reduction and Limitation of Strategic Offensive Arms, U.S.-U.S.S.R., July 31, 1991, www.state.gov/documents/organization/27360.pdf
6 Pike et al., "WMD Around the World: UR-100N/SS-19 STILLETO."
7 Missile Defense Project, "SS-19 "Stiletto," *Missile Threat*, Center for Strategic and International Studies, published August 10, 2016, last modified June 15, 2018, https://missilethreat.csis.org/missile/ss-19/.
8 Hans M. Kristensen and Robert S. Norris, "Russian Nuclear Forces, 2018," *Bulletin of the Atomic Scientists* 74, no. 3 (April 30, 2018), 185–195.
9 "Russia Launched Serial Production of Avangard Hypersonic Missile," *Army Recognition*, March 7, 2018, http://armyrecognition.com/march_2018_global_defense_security_army_news_industry/russia_launched_serial_production_of_avangard_hypersonic_missile.html.
10 John Pike et al., "WMD Around the World: R-36M/SS-18 SATAN," *Federation of Atomic Scientists*, last modified July 29, 2000, https://fas.org/nuke/guide/russia/icbm/r-36m.htm; The Soviet Union's Strategic Rocket Forces were formed in 1959 and were responsible for the operations of all nuclear warhead-armed ground-based intercontinental missiles, intermediate-range ballistic missile, and medium-range ballistic missiles.
11 Pike et al., "WMD Around the World: R-36M/SS-18 SATAN."
12 Ibid.
13 Kristensen and Norris, "Russian Nuclear Forces, 2018."
14 MIRV versions of this weapon were tested but not deployed.
15 John Pike et al., "WMD Around the World: RT-2PM – SS-25 SICKLE," *Federation of Atomic Scientists*, last modified July 29, 2000, https://fas.org/nuke/guide/russia/icbm/rt-2pm.htm.
16 Kristensen and Norris, "Russian Nuclear Forces, 2018."
17 Ibid.
18 While three warheads were carried, START I treaty considered all missile to carry four MIRVs.
19 Pike et al., "WMD Around the World: RT-2PM – SS-25 SICKLE."
20 Kristin Horitski, "R-29R/SS-N-18 Stingray," *Missile Defense Advocacy Alliance*, last modified December 15, 2015, http://missiledefenseadvocacy.org/missile-threat-and-proliferation/missile-proliferation/russia/r-29r-ss-n-18-stingray/.
21 Kristensen and Norris, "Russian Nuclear Forces, 2018," 189–190.
22 Missile Defense Project, "SS-N-18 'Stingray'," *Missile Threat*, Center for Strategic and International Studies, published August 10, 2016, last modified June 15, 2018, https://missilethreat.csis.org/missile/ss-n-18/.
23 The SS-N-23 warhead capacity is uncertain. It was reported as having been tested with 10, which aligns with U.S. intelligence estimates that the missile could carry up to 10, but this number reduced to four in 2009. Additionally, treaty obligations would restrict the number of warheads to four.
24 This aligns with: National Air and Space Intelligence Center, "Ballistic and Cruise Missile Threat," NASIC-1031-0985-13, 2013, http://fas.org/programs/ssp/nukes/nuclearweapons/NASIC2013_050813.pdf; Kristensen and Norris, "Russian Nuclear Forces, 2018," 186.
25 Kristensen and Norris, "Russian Nuclear Forces, 2018," 190.

26 John Pike et al., "WMD Around the World: Tu-95 BEAR (TUPOLEV)," *Federation of Atomic Scientists*, last modified August 8, 2000, https://fas.org/nuke/guide/russia/bomber/tu-95.htm; John Pike et al., "WMD Around the World: Kh-55 Granat, AS-15 Kent, SS-N-21 Sampson, SSC-4 Slingshot," *Federation of Atomic Scientists*, last modified August 8, 2000, https://fas.org/nuke/guide/russia/bomber/as-15.htm.
27 12 is provided by Kristensen 2018, 24 by https://fas.org/nuke/guide/russia/bomber/tu-160.htm.
28 Anton Valagin, "Все бомбардировщики Ту-160 модернизируют [All Tu-160 Bombers Are Upgrading]," *Rossiyskaya Gazeta*, April 28, 2017, https://rg.ru/2017/04/28/reg-pfo/vse-bombardirovshchiki-tu-160-moderniziruiut.html; United Aircraft Corporation, "Обновленный 'Медведь': компания «Туполев» продолжает работы по модернизации ракетоносцев Ту-95МС [Updated 'Bear': The Company 'Tupolev' continues to work on the modernization of missile carriers Tu-95MS]," December 26, 2016, https://uacrussia.livejournal.com/57594.html; Kristensen and Norris, "Russian Nuclear Forces, 2018," 190.
29 U.S. Defense Intelligence Agency, "Russia Military Power: Building a Military to Support Great Power Aspirations," DIA-11-1704-161, 2017, www.dia.mil/Military-Power-Publications.
30 Kristensen and Norris, "Russian Nuclear Forces, 2018," 190.
31 Ibid., 31.
32 Ibid., 186.
33 Pavel Podvig, "Russia's Current Nuclear Modernization and Arms Control," *Journal for Peace and Nuclear Disarmament* 1, no. 2 (2018), 256–267.
34 Gaukhar Mukhatzhanova, *Implementation of the Conclusions and Recommendations for Follow-on Actions Adopted at the 2010 NPT Review Conference Disarmament Actions 1–22: 2014 Monitoring Report* (Monterey, CA: James Martin Center for Nonproliferation Studies, 2014), 13.
35 "Putin: Russia to Deploy New Weapons to Counter U.S. Missile Shield," *CBS/AP*, November 10, 2015, www.cbsnews.com/news/putin-russia-will-deploy-new-weapons-to-counter-natos-missile-shield/.
36 U.S. Department of Defense, Office of the Secretary of Defense, "Nuclear Posture Review 2018."
37 As an aside, this period looks more like the mid-1980s, when nuclear tensions were increasing, new technologies brought online, and rapid development of increasing number of warheads was occurring.
38 Kristensen and Norris, "Russian Nuclear Forces, 2018," 186.
39 "РС-28/ОКР Сармат, ракета 15А28 – SS-X-30 SARMAT [RS-28/OKR Sarmat, Missile 15A28 – SS-X-30 SARMAT]," *Military Russia Blog*, March 31, 2018, http://militaryrussia.ru/blog/topic-435.html; Pavel Podvig, "Yes, Sarmat Program Is in Trouble," *Russian Strategic Nuclear Forces Blog*, July 3, 2017, http://russianforces.org/blog/2017/07/yes_sarmat_program_is_in_troub.shtml; Pavel Podvig, "Sarmat Ejection Test, at Last," *Russian Strategic Nuclear Forces Blog*, December 29, 2017, http://russianforces.org/blog/2017/12/sarmat_ejection_test_at_last.shtml.
40 "Putin: Russia Creates Advanced Weapons Responding to U.S. Scrapping Missile Treaty," *Sputnik International News*, March 1, 2018, https://sputniknews.com/russia/201803011062108691-putin-russia-weapons-us-missile-treaty/.
41 "Objekt 4202/Yu-71/Yu-74," *GlobalSecurity.org*, last updated March 4, 2018, www.globalsecurity.org/wmd/world/russia/objekt-4202.htm; "Борисов: испытания комплекса 'Авангард' доказали его способность разгоняться до 27 Махов [Borisov: Tests of the Avangard Complex Proved Its Ability to Accelerate to 27 Machs]," *TASS*, December 27, 2018, https://tass.ru/armiya-i-opk/5958896.
42 "Russia Excludes Rail-Mobile ICBM System from Armament, Focuses on Sarmat Missile," *TASS*, December 6, 2017, http://tass.com/defense/979334.

43 Pavel Podvig, "RS-26 and Other Intermediate-Range ICBMs," *Russian Strategic Nuclear Forces Blog*, July 18, 2017, http://russianforces.org/blog/2017/07/rs-26_and_other_intermediate-r.shtml; Alexey Nikolski, "Госпрограмму вооружений поправят с учетом представленной Путиным ракеты [The State Armament Program Will Be Corrected in View of the Missile Presented by Putin]," *Vedomosti*, March 22, 2018, www.vedomosti.ru/politics/articles/2018/03/22/754613-gosprogrammu-vooruzhenii-popravyat.

44 Justin Mohn, "Russia's Submarine Force Is Getting Stronger. How Worried Should We Be?" *The National Interest*, July 7, 2018, https://nationalinterest.org/blog/buzz/russias-submarine-force-getting-stronger-how-worried-should-we-be-25172.

45 Hans M. Kristensen and Robert S. Norris, "Russian Nuclear Forces, 2017," *Bulletin of the Atomic Scientists* 73, no. 2 (February 2017), 115–126; "Россия начнет строить новые АПЛ «Борей» в 2012 году [Russia to Begin Construction of New Borey-Class Nuclear Submarines in 2012]," *RIA Novosti*, December 2, 2011, https://ria.ru/20111202/504183548.html; "Alexander Nevsky Submarine to Join Russian Navy on Nov 15," *TASS*, July 6, 2013, http://tass.com/russia/696640; "Russia's Third Borey Class Sub Blessed for Sea Trials," *RIA Novosti*, June 10, 2013, https://sptnkne.ws/kJX9; Pavel Podvig, "Strategic Fleet," *Russian Strategic Nuclear Forces Blog*, June 26, 2017, http://russianforces.org/navy/.

46 Pavel Podvig, "Russia to Build Six More Borey-A Submarines," *Russian Strategic Nuclear Forces Blog*, May 26, 2018, http://russianforces.org/blog/2018/05/russia_to_build_six_more_borey.shtml.

47 Pavel Podvig, "Bulava Missile Test History," *Russian Strategic Nuclear Forces Blog*, November 28, 2014, http://russianforces.org/blog/2007/11/bulava_test_history.shtml.

48 Pavel Podvig, "Bulava Is Finally Accepted for Service," *Russian Strategic Nuclear Forces Blog*, June 29, 2018, http://russianforces.org/blog/2018/06/bulava_is_finally_accepted_for.shtml.

49 Charlie Gao, "Russia's Husky Class Submarine: Armed with Nuclear Torpedoes and Hypersonic Missiles?" *The National Interest*, May 10, 2018, https://nationalinterest.org/blog/the-buzz/russias-husky-class-submarine-armed-nuclear-torpedoes-25784.

50 Pavel Podvig, "First Tu-160M2 Takes Flight, Production Contract for Ten Aircraft Signed," *Russian Strategic Nuclear Forces Blog*, January 26, 2018, http://russianforces.org/blog/2018/01/first_tu-160m2_takes_flight_pr.shtml; "Секреты ПАК ДА и Ту-160М2: каким будет крылатый щит России [Secrets of the PAK-DA and Tu-160M2: What Will Be the Winged Shield of Russia]," *Tass*, December 22, 2017, https://tass.ru/armiya-i-opk/4827520.

51 Jerry Davydov and Bryan Lee, "Russia's Nuclear Rearmament: Policy Shift or Business as Usual," *NTI Issue Brief*, December 18, 2013, www.nti.org/analysis/articles/russias_nuclear_rearmament_policy_shift_or_business_as_usual/.

52 Pavel Podvig, "Plans for the New Strategic Bomber," *Russian Strategic Nuclear Forces Blog*, May 22, 2014, http://russianforces.org/blog/2014/05/plans_for_the_new_strategic_bo.shtml; "Russian Space-Age Stealth Bomber to Hit the Skies with Hypersonic Weapons," *Sputnik International News*, April 20, 2016, https://sputniknews.com/russia/201604201038309843-russia-stealth-bomber/.

53 Kristensen and Norris, "Russian Nuclear Forces, 2018," 190.

54 "Remarks by Michael R. Pompeo, Secretary of State on INF Treaty," U.S. Department of State, February 1, 2019, https://fr.usembassy.gov/remarks-by-michael-r-pompeo-secretary-of-state-on-inf-treaty/.

55 Andrew E. Kramer, "Russia Pulls Out of I.N.F. Treaty in 'Symmetrical' Response to U.S. Move," *The New York Times*, February 2, 2019, www.nytimes.com/2019/02/02/world/europe/russia-inf-treaty.html.

56 Lisa Schlein, "Time Running Out for Salvaging INF Treaty," *VOA News*, March 27, 2019, www.voanews.com/a/time-running-out-for-salvaging-inf-treaty/4850703.html.

57 U.S. Department of State, "INF Diplomatic Timeline," Fact Sheet, Washington, DC, February 1, 2019, www.state.gov/t/avc/inf/287411.htm.

58 Frank A. Rose, "The End of an Era? The INF Treaty, New START, and the Future of Strategic Stability," *Brookings' Order from Chaos*, February 12, 2019, www.brookings.edu/blog/order-from-chaos/2019/02/12/the-end-of-an-era-the-inf-treaty-new-start-and-the-future-of-strategic-stability/.

59 "Director of National Intelligence Daniel Coats on Russia's INF Treaty Violation," Office of the Director of National Intelligence, November 30, 2018, www.dni.gov/index.php/newsroom/speeches-interviews/item/1923-director-of-national-intelligence-daniel-coats-on-russia-s-inf-treaty-violation.

60 Idrees Ali and Arshad Mohammed, "U.S. to Start Making Parts for Ground-Launched Cruise Missile Systems," CSR *Times*, March 12, 2019, www.thecsrtimes.com/2019/03/12/u-s-to-begin-making-elements-for-ground-launched-cruise-missile-programs/.

61 Anton Troianovski, "Putin Claims Russia Is Developing Nuclear Arms Capable of Avoiding Missile Defenses," *The Washington Post*, March 1, 2018, www.washington-post.com/world/europe/putin-claims-russia-has-nuclear-arsenal-capable-of-avoiding-missile-defenses/2018/03/01/d2dcf522-1d3b-11e8-b2d9-08e748f892c0_story.html?utm_term=.44ce12d7016f.

62 "Objekt 4202/Yu-71/Yu-74"; "[Borisov: Tests of the Avangard Complex Proved Its Ability to Accelerate to 27 Machs]."

63 "'Listen to Us Now': Putin Unveils New Russian Nuclear Arsenal," *Russia Today*, published March 1, 2018, last updated July 19, 2018, www.rt.com/news/420206-russia-strategic-weapons-putin/.

64 U.S. Defense Intelligence Agency, "Russia Military Power."

65 Troianovski, "Putin Claims Russia Is Developing Nuclear Arms Capable of Avoiding Missile Defenses."

66 Aaron Mehta, "One Nuclear Treaty Is Dead. Is New START Next?" *Navy Times*, October 23, 2018, www.navytimes.com/pentagon/2018/10/23/one-nuclear-treaty-is-dead-is-new-start-next/.

4 U.S.-Russian bilateral disarmament

Cameron Trainer

The modernization of the Russian Federation's nuclear armaments has been underway for quite some time. As a country, like the United States, with a developed military-industrial complex, its efforts to develop new or improved armaments never really cease. This certainly holds true for its nuclear weapons. Nonetheless, it is possible to situate Russia's current efforts to modernize its nuclear forces in the context of the country's push, from the early 2000s on, to revitalize the Russian armed forces. In this sense, Russia's modernization campaign was initiated for domestic reasons. Nonetheless, Russia's modernization has informed – and, in turn, been informed by – the military development of other nations.

True, the turn of the millennium is sometimes considered to have been ushered in along with a new era of multipolarity. This has certainly informed U.S. thinking. President Barack Obama notably embraced a 'pivot to Asia'. But, as will be demonstrated throughout this chapter, Russia has consistently characterized its nuclear forces vis-a-vis those of the United States. And, particularly as relations between the two countries have worsened, the United States has once again taken to using Russia's nuclear capabilities to inform its own investments. This would imply that the dynamic between the two countries – at least where nuclear armaments are concerned – best fits the action-reaction model.

The action-reaction model, as outlined by Barry Buzan and Eric Herring in their 1998 book *The Arms Dynamic in World Politics*, is centered on the assertion "that states strengthen their armaments because of the threats the states perceive from other states."[1] Buzan and Herring further state that, under this model, an action by a state will prompt reciprocal action from its potential adversaries,[2] hence, the "action-reaction" moniker.

Scholars in both the United States and Russia skeptical of the relevance of the action-reaction model commonly point to domestic factors in each country. These scholars are quick to point out the influence, particularly in Russia, wielded by the military-industrial complex.[3] Domestic politics are also taken to be a factor, especially with regard to constraining the options available to the American president.[4] Yet, these domestic drivers quite often leverage the language of international competition to achieve their domestic aims. That this is done in both the United States and Russia with specific reference to the other sufficiently establishes the applicability of the action-reaction model. The model, after all, places no constraints on the motives according to which states make their decisions to arm, requiring only that the state feel threatened by an external actor.[5]

Irrespective of its origin, the dynamic between the United States and Russia – at least in terms of nuclear modernization – meets the criteria of the action-reaction model as detailed throughout this chapter. Moreover, each country's modernization programs outpace, or are otherwise apart from the technological imperative, the need to keep pace with the qualitative advancement of technology.[6]

This chapter will examine the action-reaction dynamics of Russian and American nuclear modernization in three parts. The first will focus on strategic nuclear weapons, with the discussion centered on the classic components of the nuclear triad – ground-, air-, and sea-launched delivery systems affixed with nuclear warheads – as well as the impact that missile defense programs have had on other investments in strategic capabilities. The second will take aim at non-strategic nuclear weapons and the ways that the United States and Russia envision using them. The debate over the existence of a Russian 'escalate to de-escalate' doctrine will also be discussed in this section. Finally, the third part will provide a brief look at the impact of arms reduction and control on the dynamics of nuclear arms modernization.

As this chapter is on Russian nuclear modernization, it will often pull from Russian sources, both governmental and nongovernmental. Because the action-reaction model places a premium on how a country *perceives* the actions of others rather than the intent of those actions or their actual effects, this chapter references many public statements by government actors. Such statements may not be objective relations of fact. Their accuracy or lack thereof is based not only on the knowledge of the actor but also upon their motives, i.e. whether distortion of the truth may better serve their political interests. Similarly, public remarks – by their very nature – do not capture information that the United States or Russia have classified. This chapter, therefore, is based on an incomplete and likely imperfect understanding of the actual modernization efforts underway in both the United States and Russia.

Strategic nuclear forces

The strategic nuclear forces of the Russian Federation, like those of the United States, are composed of ground-launched intercontinental ballistic missiles (ICBMs), long-range bombers, and submarine-launched ballistic missiles (SLBMs). Russian doctrine entrusts these components – the nuclear triad – with the prevention of a nuclear conflict or any other military conflict. Language to this extent is included in all versions of the Military Doctrine of the Russian Federation (first issued in 1993; most recently issued in 2014), as well as Russia's most recent National Security Strategy.[7] An earlier iteration of Russia's National Security Strategy, then termed the National Security Concept, specified that this task was predicated upon the strategic nuclear forces being "capable of assuredly inflicting the desired extent of damage against any aggressor state or coalition of states in any conditions and circumstances."[8] This firmly situates Russia's strategic nuclear forces in a deterrence role.

Russia's strategic nuclear forces have ably fulfilled the deterrence mission for decades. Even in 1998, when Russia's armed forces were at their most resource-constrained,[9] Russia's strategic nuclear forces remained a credible deterrent. The sheer size of the Russian arsenal – then estimated at near 6,000 warheads spread over some 1,100 launchers[10] – all but assured Russian existential security. Nonetheless, Russia's strategic nuclear forces did face serious issues due to a lack of funds. Incidents like the sinking of the *Kursk*, a Russian nuclear-powered cruise-missile submarine (not part of Russia's strategic nuclear forces, though capable of being armed with non-strategic nuclear weapons),[11] meant that the challenges faced by Russia's armed forces were far from private. The degradation of the submarine and bomber legs of Russia's nuclear triad even led some observers to question whether Russia would have to rely solely on a land-based nuclear deterrent.[12] Reliance only on the Strategic Rocket Forces, the branch of the Russian armed forces that comprises the 'land leg' of the nuclear triad, would have increased Russia's vulnerability to a disarming (counterforce) strike targeting its nuclear arsenal.

To be sure, a successful counterforce strike against Russia's land-based strategic nuclear weapons would have been a tall order. Russia, in 1998, possessed over 750 ICBMs accounting for over 3,500 nuclear warheads and spread across 19 main operating bases, with almost 400 of the ICBMs being either road- or rail-mobile.[13] Yet Russia did not – and does not, as exemplified by its continued commitment to the land, sea, and air legs of its nuclear triad – appear willing to risk even such an unlikely event as a disarming strike neutralizing its land-based nuclear deterrent.

Concern over potential vulnerability to a counterforce or disarming strike was exacerbated by the unilateral American withdrawal from the 1972 Anti-Ballistic Missile (ABM) Treaty in 2002. American missile defense programs – particularly the deployment of ballistic missile defense systems on Russia's periphery – have been one of the persistent stressors of U.S.-Russia relations.

For the Russian side, American missile defense programs are taken as a potential threat to the survivability of Russia's nuclear deterrent. Russia does not necessarily view current missile defense capabilities as a serious threat. Rather, Russia fears – or at least publicly claims to fear – that missile defense technologies might mature to the point that the United States would be able to sufficiently blunt the impact of a Russian second strike, thereby nullifying Russia's nuclear deterrent.[14] This is somewhat supported by the principles of missile defense as outlined in the United States' 2002 *Nuclear Posture Review*, which indicate that the United States "seeks effective defenses against attacks by small numbers of longer range missiles."[15] This could be interpreted to include any Russian missiles surviving an American first strike. Russian President Vladimir Putin, speaking at the 2007 Munich Security Conference, worried that this might lead the United States to use military force with impunity "not only in local but eventually also in global conflicts."[16] In the years following, Russian officials have continued to make similar statements in times of relative rapprochement[17] and tension[18] with the United States. Part of this may have to do with Russian domestic political thought. Writing in 2008, Professor Stephen Blank of the U.S. Army War College linked

Russia's reaction to the U.S. withdrawal from the ABM Treaty to the "perception of Russia as a besieged fortress and the primary global counterpole to America and the West."[19] This helps to explain why Russia's modernization of its strategic nuclear forces appears to have been consistently driven by a perceived technological imperative to counter future missile defense systems.

But from a more cynical perspective, Russia's comments on the potentially destabilizing impact of American missile defense systems may be interpreted as a *post hoc* justification for the modernization of its nuclear forces. This would appear to be supported by Putin's 2001 statement on the American decision to unilaterally withdraw from the ABM Treaty, in which he claimed that Russia already possessed the capabilities needed to penetrate any ABM defenses and that, consequently, they were "not a serious threat for Russia."[20] Scholars embracing this perspective have pointed out that Russia has been willing to overlook the issues they claim are generated by missile defense programs when politically expedient. Director of the Russian Nuclear Forces Project Pavel Podvig, for instance, identified the thaw in U.S.-Russia relations after 9/11 and the signing of the New Strategic Arms Reduction Treaty (New START) as cases where other interests prevailed over the issue of missile defense.[21] That other priorities in some cases prevail over missile defense concerns should, however, not be taken as evidence that Russia is less than serious about countering the effects of American ballistic missile defense. Indeed, the choices made by Russia in modernizing each of the three components of its strategic nuclear forces seem expressly designed to hedge against improvements in missile defense technologies.

Strategic rocket forces

The Strategic Rocket Forces currently account for an estimated 1,165 of the approximately 1,600 nuclear warheads associated with Russia's strategic nuclear weapons systems.[22] These systems – all ICBMs – include the SS-18 (RS-20V), SS-19 (RS-18 or UR-100NUTTH), SS-25 (RS-12M Topol), and SS-27 Mods 1 and 2 (RS-12M1/M2 Topol-M and RS-24 Yars). The SS-18, SS-19, and SS-25 are all remnants of the Soviet nuclear arsenal and are expected to be decommissioned in the early 2020s. The SS-27 Mods 1 and 2, the only ICBMs first deployed after the fall of the Soviet Union, now make up a majority of Russia's ICBM arsenal.

Russia considers the SS-27 Mods 1 and 2 to be separate systems (hence the differing names, Topol-M and Yars).[23] The main difference between the two, however, is the use of multiple independent re-entry vehicles (MIRV) on the Yars. The Yars, first deployed in 2010, is the current focus of Russia's ICBM modernization efforts.[24] The shift from one warhead per ICBM (the Topol-M) to multiple independently targetable re-entry vehicle (MIRVed) ICBMs (the Yars) can be interpreted as a response to the United States' commitment to missile defense systems and Russia's perception that this increases the risk of an American counterforce strike. This is because the multiple warheads of MIRVed missiles are understood to overwhelm missile defense systems. Indeed, Russia framed the Yars as a reaction to the expansion of American missile defenses, trumpeting the MIRVed system as 'almost impossible to shoot down' following a test of the system in 2007.[25] The next stage of Russia's ICBM modernization, the planned deployment of the SS-X-30 (RS-28 Sarmat), looks to be a continuation of Russia's reaction to American missile defenses. The Sarmat will be MIRVed with an estimated ten plus warheads, possibly incorporating the Avangard hypersonic glide vehicle to evade rather than solely overwhelm missile defenses.[26] *Sputnik*, a news agency controlled by the Russian state-owned Rossiya Segodnya, has explicitly framed the Sarmat as a response to both the expansion of American missile defenses and the development of conventional capabilities like Prompt Global Strike.[27]

Strategic fleet

Russia's fleet of nuclear-powered ballistic missile submarines (SSBNs) is comprised of ten vessels, each accounting for 16 submarine-launched ballistic missiles (SLBM), in three classes: six Delta IV (Project 667BRDM), one Delta III (Project 667BRD), and three Borei (Project 955).[28] The Borei-class SSBNs are the modernizing component of the naval leg of Russia's triad.

Construction began on the first Borei-class SSBN in 1996, but was temporarily suspended to allow for a redesign incorporating the new SS-N-32 (RSM-56 Bulava) SLBM with an expected commissioning of the first vessel in 2005.[29] The first Borei-class SSBN was actually commissioned in 2013 – eight years later than anticipated.[30] The three operational Borei SSBNs are to be complemented by at least five additional submarines designated class Borei-A, the first of which was initially slated for delivery in 2017 but is now expected to be delivered in 2019.[31] Reporting by the Russia *TASS* news agency indicates that Russia seeks to acquire a total of 11 Borei-A SSBNs, which – when accounting for the planned retirement of the Delta III and Delta IV SSBNs – would increase Russia's SSBN fleet to 14.[32] Note too that the Bulava SLBM has the potential to carry six MIRVed warheads, more than either the SS-N-18 (RSM-50) or the SS-N-23 (RSM-54 Sineva) assigned to Delta III and IV SSBNs.[33]

Because of the new Borei- and Borei-A-class SSBNS and reductions mandated by arms control treaties, most notably the New Strategic Arms Reduction Treaty signed with the United States in 2010, Russia's ballistic missile submarines in 2017 accounted for a greater proportion (31%) of the country's nuclear warheads than at any point since the turn of the millennia.[34] If the planned transition to an all-Borei SSBN fleet of 14 vessels were to occur (and the number of warheads assigned to ICBMs and long-range bombers remained constant), approximately 41% of Russia's strategic nuclear warheads would be assigned to its naval systems. This is particularly notable given that the proportion of warheads assigned to SSBNs dropped to a low of 18% in 2005.[35] This would appear to indicate – particularly in conjunction with the Bulava SLBM's greater warhead capacity – that Russia places a premium on maintaining a strategic nuclear force capable of surviving a potential American first strike. As previously detailed, this could be taken as a response to a perceived increase in threat created in part by the expansion of American missile defenses.

Strategic aviation

Russia's strategic aviation capabilities are currently provided by two nuclear-capable heavy bombers, the Tu-160 and the Tu-95MS, both of which are undergoing modernization efforts that include equipping the bombers to field the AS-23B (Kh-102, in Russia) air-launched cruise missile (ALCM) in lieu of the Soviet-era AS-15 (Kh-55).[36] The Kh-102 is seen as giving "Russia's bomber fleet a long-range precision strike capability that was until recently the sole purview of the U.S. Air Force."[37] The system is, per the Center for Strategic and International Studies, "designed to defeat air defense systems by flying at low, terrain-hugging altitudes to avoid radar systems."[38] Russia is also looking forward to replacing its strategic bombers: it anticipates introducing a variant of the Tu-160, termed the Tu-160M2, in 2023 and an entirely new long-range bomber – currently termed the PAK DA – at an indeterminate point in the future.[39] The PAK DA will incorporate stealth capabilities akin to those planned for the PAK FA, Russia's response to the American F-22.[40] These advances, most notably the design (and rhetorical framing) of the Kh-102 in relation to American capabilities, continue to demonstrate Russia's development of its strategic nuclear forces vis-à-vis American capabilities.

In addition to modernizing all aspects of its nuclear triad, Russia is developing 'exotic' weapons systems that defy classification as part of any particular leg of the traditional triad, namely a ground-launched nuclear-powered cruise missile (Burevestnik) and a submarine-launched drone (Poseidon).[41] Russian President Vladimir Putin framed, in remarkably clear terms, the Burevestnik and Poseidon systems – in addition to the Sarmat ICBM – as a response to American missile defenses, stating: "How will Russia respond to this [missile defense] challenge? This is how."[42]

Russia gives the appearance of taking the development of these systems seriously. Addressing the Russian Federal Assembly, Putin touted that the systems were "successfully undergoing tests"[43] and the Poseidon, at least, appears to be moving closer to eventual deployment. Russia launched a submarine – a repurposed Oscar II class vessel separate from the Borei- and Delta-class submarines that comprise Russia's strategic fleet – to serve as a platform for the system in April 2019.[44] This matches with reporting from January 2019 by Russian news source, *TASS*, anticipating that up to 32 "strategic" Poseidon drones could be deployed across four submarines.[45]

Despite Russia's ambitious plans, the modernization of all aspects of Russia's strategic nuclear programs have suffered incidents and setbacks. The novel Burevestnik cruise missile appears to be facing 'developmental challenges'.[46] The development of some programs, like the Rubezh ICBM and the Barguzin rail-mobile ballistic missile system, has been suspended.[47] And, as detailed above, the introduction of the first Borei SSBN has been delayed eight years. That these programs have had challenging developments, however, should not be seen as a lack of commitment to their modernization. In fact, over the course of this decade, Russia's nuclear weapons spending increased as a proportion of its total spending on national defense.[48] This, along with the concentration on systems not susceptible to existing or anticipated missile defense capabilities, clearly demonstrates the priority Russia has placed on its nuclear arms modernization. And it is particularly significant considering its nuclear forces have quantitatively decreased so as to meet the requirements of the New Strategic Arms Reduction Treaty.

American modernization efforts (though continuous in the same sense as those of Russia) are generally taken to have been prioritized after those of Russia. This is perhaps most clearly seen through the differences between the *Nuclear Posture Reviews* released by the United States in 2002, 2010, and 2018.

The 2002 *Nuclear Posture Review* elevated defensive capabilities, missile defenses being most prominently featured, to what it termed the 'New Triad' in recognition that non-state or rogue actors might not be deterred by the offensive triad of ICBMs, long-range bombers, and SLBMs. This, as detailed above, had the unintended consequence of stoking Russian anxieties and driving the development of systems that can evade rather than simply overwhelm missile defense systems.

The 2010 *Nuclear Posture Review* focused on maintaining the offensive triad at the reduced force levels, though it also directed the United States Navy to begin technology development of a new SSBN[49] and noted that the United States would examine ICBM[50] and ALCM[51] replacements. It named Russia as the only country with a nuclear arsenal comparable to that of the United States, listing Russia's nuclear forces as a 'significant factor' in determining the size of U.S. nuclear forces.[52] This demonstrates the enduring focus on Russia as a potential adversary, despite the mutual arms reductions mandated by the New Strategic Arms Reduction Treaty and 2010's relatively good relations between the United States and Russia. This characterization of Russia as an enduring focus of American policymaking is further supported by the 2010 *Nuclear Posture Review's* acknowledgment that "large disparities in nuclear capabilities could raise concerns on both sides and among U.S. allies and partners, and may not be conducive to maintaining a stable, long-term strategic relationship, especially as nuclear forces are significantly reduced."[53] While this discussion of capability was framed in quantitative terms, the same holds true for qualitative differences in capability.

The 2018 *Nuclear Posture Review*, with its focus on replacing existing systems, addressed the potential for a qualitative gap in nuclear capability. Many of the proposed replacements can be seen as extensions of the measures contained in the 2010 *Nuclear Posture Review*. These include the Columbia-class SSBN, the new ICBM (coined the 'Ground-Based Strategic Deterrent'), and the new ALCM (the 'long-range standoff' or 'LRSO' cruise missile). Importantly, these programs are all couched in terminology showing a justification apart from any external actor. The new SSBN and ICBM replacements are deemed necessary because the service life of existing systems cannot be extended any longer.[54] The new ALCM is justified on the basis that the existing system's service life can only be extended until the new system can be made available.[55] This would imply that the United States' decisions on strategic nuclear weapons are not affected by an action-reaction dynamic with Russia. Still, the 2018 *Nuclear Posture Review* did "put a premium on the survivability, flexibility and readiness of Western nuclear and non-nuclear capabilities to hold diverse types of Russian targets at risk throughout a crisis or conflict."[56] This, at the very least, suggests that there is criteria by which Russia's development of its strategic nuclear forces would elicit a clearer American response. Indeed, this would match the American response to Russia's continued development of its nuclear-capable Non-strategic weapons, as examined in the following section.

Non-strategic nuclear weapons, doctrine

While strategic nuclear forces play a prominent role as a deterrent for both Russia and the United States, the role of non-strategic nuclear systems is less clearly defined. During the Cold War, non-strategic nuclear weapons were envisioned as battlefield weapons to counter superior conventional forces.[57] At the height of the Cold War, the United States maintained more than 7,000 operational non-strategic nuclear warheads while, at the end of the Cold War, the Soviet Union may have held an arsenal of over 20,000 non-strategic nuclear weapons.[58] Since then, the arsenals of non-strategic nuclear weapons maintained by Russia and the United States have decreased substantially. Today, it is estimated that the United States retains 200 non-strategic nuclear warheads to Russia's 1,830.[59] This represents a drastic decrease from the Cold War era. Moreover, Russia has placed its non-strategic nuclear weapons in a non-deployed status, which Russian officials have characterized as "the biggest de-alerting measure in world history."[60] The United States, for its part, keeps 150 non-strategic nuclear weapons forward-deployed in Europe, while the remainder are kept in central storage in the United States.[61] Despite the decreased emphasis on non-strategic nuclear weapons systems, both the United States and Russia are modifying or replacing existing systems.

The United States is, per the 2018 *Nuclear Posture Review*, planning to "modify a small number of existing SLBM warheads to provide a low-yield option" and "pursue a modern nuclear-armed sea-launched cruise missile (SLCM)."[62] While the low-yield SLBM is a modification of an existing system and therefore not a revolutionary new capability, the planned nuclear-armed SLCM would fill a gap in the U.S. arsenal. But that gap was left when the previous nuclear-armed SLCM's retirement was announced in the 2010 *Nuclear Posture Review*,[63] so it too should not be interpreted as a radical re-envisioning of U.S. nuclear policy. Complementing these newly planned capabilities is the shift to the B61-12 gravity bomb (from earlier versions of the B61) for the forward-deployed F-15E dual-capable aircraft.[64] The F-15E will also be replaced in its role of dual-capable aircraft by the F-35.[65] Note that, while these plans are explicitly laid out in the 2018 *Nuclear Posture Review* and are apparent priorities of the Trump administration, elements of the U.S. Congress have fiercely contested the deployment of low-yield nuclear capabilities to the U.S. Navy.[66] How this will ultimately play out is, as of yet, unknown.

Russia, meanwhile, possesses non-strategic nuclear weapons – as defined by the United States' 2018 *Nuclear Posture Review* – including,

> air-to-surface missiles, short range ballistic missiles, gravity bombs, and depth charges for medium-range bombers, tactical bombers, and naval aviation, as well as anti-ship, anti-submarine, and anti-aircraft missiles and torpedoes for surface ships and submarines, a nuclear ground-launched cruise missile in violation of the 1987 INF Treaty, and Moscow's antiballistic missile system.[67]

Two systems are especially notable. The first is the "nuclear ground-launched cruise missile in violation of the 1987 INF Treaty."[68] It is identified by the United States as the SSC-8 (Russian designation: Novator 9M729) and prompted the United States to begin research and development of its own ground-launched intermediate-range missile.[69] The second is the SS-26 (Iskander-M, in Russia) short-range ballistic missile. Its deployments to Kaliningrad have been framed as fulfilling Russian promises to counter the deployment of ballistic missile defenses in Europe.[70] In this respect, it is a successor to the SS-21 (OTR-21 Tochka), which was the system previously rumored to be Russia's nuclear capability in Kaliningrad.[71] Many – if not all – of Russia's non-strategic nuclear systems are dual-capable, meaning they can be armed with either conventional or nuclear warheads. This has resulted in significant ambiguity in the role Russia envisions these systems being used in and whether this is representative of an increased willingness to use nuclear weapons.

The 2018 *Nuclear Posture Review* asserts that Russia is "increasing the total number of such [non-strategic nuclear] weapons in its arsenal, while significantly improving its delivery capabilities."[72] This is fairly disingenuous. Russia's arsenal of non-strategic nuclear warheads has decreased not only in relation to Soviet stockpiles as noted above, but also in comparison to those of Russia at the turn of the decade.[73] The *Nuclear Posture Review* does accurately capture the improvement of Russia's delivery capabilities and can be interpreted as describing the increasing number of those delivery systems rather than the number of non-strategic nuclear warheads. But these systems also serve – and have been actually used as – conventional weapons. The Iskander-M is the best example of this. Not only has it been used in a conventional role in Syria,[74] but it was not even initially conceptualized by Russia as filling a nuclear role.[75] For this reason, increased focus on these systems does not necessarily indicate any change in Russia's approach to non-strategic nuclear weapons.

The continued development of non-strategic weapons by the United States and Russia has become a flashpoint in large part due to how each country perceives the other as being prepared to use them. This has resulted in non-strategic nuclear weapons being one of the clearest examples of the action-reaction dynamic at play. This is largely due to American interpretation of Russian doctrine as endorsing an 'escalate to de-escalate' strategy in which non-strategic nuclear weapons could be used to force a favorable outcome if Russia began to lose a conventional war with NATO.[76]

The existence of a Russian 'escalate to de-escalate' strategy has been met with skepticism by some scholars. The Center for Strategic and International Studies' Olga Oliker, for example, described Russian behavior as indicating "less a lowering of the threshold [for nuclear force] than a reminder that escalation is possible, and that Russia must therefore be taken seriously."[77] Others, like Hans M. Kristensen and Robert S. Norris, point to the scarcity of public evidence of such a strategy's existence.[78] Still other experts point out that non-strategic nuclear weapons are not appointed a role in Russia's military doctrine, seeing this as evidence that Russia itself questions the utility of such weapons.[79]

The United States government, however, appears to have accepted the existence of an 'escalate to de-escalate' strategy as fact. This is most evident in the decision to address such a strategy in the 2018 *Nuclear Posture Review*, which contained specific language to push back against the idea that limited nuclear use would compel the United States and its allies to de-escalate a conflict to Russia's favor.[80]

The notion that Russia has a 'escalate to de-escalate' strategy is hardly new. It has been articulated by experts on Russian nuclear policy since at least 2014[81] and current government documents (the Defense Intelligence Agency's 2018 Global Nuclear Landscape, for example) trace its origin back to a 1999 article in the Russian journal *Voennaya Mysl*.[82] Yet the view holding 'escalate to de-escalate' as a feature of Russian nuclear strategy has only relatively recently gained a foothold in government policy documents. It is noticeably absent from the 2010 *Nuclear Posture Review* and alluded to in the 2014 *Quadrennial Defense Review*.[83]

American acceptance of the existence of an 'escalate to de-escalate' doctrine has been coupled with additional investment in non-strategic nuclear weapons by the United States government. This includes the decisions, noted above, to modify existing SLBMs for a low-yield option and to acquire a new SLCM. The 2018 *Nuclear Posture Review* frames these decisions in the context of an interpretation of Russia as perceiving a 'coercive advantage' in its greater number and variety of non-strategic nuclear weapons, making this a perfect example of the action-reaction dynamic at work. This is particularly true in light of the 2010 *Nuclear Posture Review*'s assessment that the SLCM the United States then possessed could be decommissioned as its deterrence and assurance roles could be "adequately substituted by these other means."[84]

The acquisition of a new SLCM by the United States is also a prime example of how action-reaction decisions can needlessly ratchet up tensions between countries. The new SLCM appears to serve no discernable purpose other than achieving capabilities parallel to those of Russia. The development of such non-strategic weapons is considered, as noted in the 2018 *Nuclear Posture Review*, "important for the preservation of credible deterrence against regional aggression."[85] But this disregards the role of conventional forces in deterring Russia from using military – including nuclear – force. Indeed, as Olga Oliker has noted, Russia is deterred "first and foremost by the threat that the United States and its allies would use their conventional supremacy to do great damage to Russian forces and preclude their military victory."[86] The 2018 *Nuclear Posture Review* tacitly acknowledges as much, stating that "U.S. strategy does not require non-strategic nuclear capabilities that quantitatively match or mimic Russia's more expansive arsenal."[87] As such, the acquisition of non-strategic nuclear systems like the new SLCM should be seen as entirely apart from the technological imperative; they serve no role that cannot be filled by other systems already possessed by the United States.

Russia, for its part, publicly asserts that such developments are based on willful misunderstanding of its policies. A statement to this effect was issued by Russian Minister of Foreign Affairs Sergei Lavrov following remarks by then-United States Secretary of Defense Ashton Carter at Minot Air Force Base in September 2016. Carter characterized potential American adversaries – clearly Russia, in context – as viewing nuclear escalation as an insurance policy backing conventional aggression,[88] which Lavrov pushed back against by promising 'reciprocal measures' in light of Carter's 'bellicose' and 'Russophobic' comments.[89] A similar statement followed the release of the United States' 2018 *Nuclear Posture Review,* which – as detailed above – declared the intent to acquire a new nuclear-equipped SLCM and modify SLBM warheads for low-yield options. In the statement, Russia denounced those systems as being 'battlefield weapons', accused the United States of lowering the nuclear threshold, and vowed to take appropriate steps to bolster national security.[90] Such steps could entail assigning non-strategic nuclear weapons a formal place in Russia's military doctrine, slowing the decommissioning of non-deployed warheads, or moving those warheads from their central storage location. The actual danger of the modernization of non-strategic nuclear weapons, therefore, has less to do with actual qualitative improvements than with the changes in their perceived roles in military doctrine.

Russia's non-strategic nuclear weapons have generally been seen as an effort to compensate for a perceived conventional weakness opposite NATO forces. This is the view articulated by Hans M. Kristensen in *SIPRI Yearbook 2018: Armaments, Disarmament and International Security*.[91] Kristensen also argues that at least some non-strategic nuclear weapons may be replaced by advanced conventional systems as those capabilities mature.[92] The danger here is that the American perception of a threat – the 'escalate to de-escalate' strategy – drives action that Russia then takes concrete steps to respond to, increasing the threat to the United States. The U.S. acquisition of a new nuclear-equipped SLCM could drive a shift, by Russia, to viewing non-strategic nuclear weapons as offsetting both a perceived conventional superiority and a nuclear threat. This would likely halt a Russian transition away from non-strategic nuclear weapons that might otherwise occur. In this manner, the NATO focus on Russia's non-strategic nuclear weapons – whose primary role, as far as there is one, is to offset NATO's conventional capabilities – could actually be a major impediment to a mutual turn away from non-strategic nuclear weapons.

Arms control and quantitative changes to the arsenals

Even as Russia and the United States modernize their strategic and tactical nuclear weapons, both countries have endorsed – at least at times – the idea of a world rid of nuclear armaments. Russian Minister of Foreign Affairs Sergei Lavrov most recently stated that Russia "remains committed to the goal of ridding the world of the nuclear threat."[93] On the American side, then-President Barack Obama notably embraced a "world without nuclear weapons" in his 2009 speech in Prague.[94] The *Nuclear Posture Reviews* published by both the Obama administration in 2010 and the Trump administration in 2018 repeated this commitment to future disarmament.

These stated commitments have periodically been reflected in arms control, limitation, and reduction measures that have both qualitatively and quantitatively limited the nuclear arsenals of both Russia and the United States. These include – among others – the 1972 Anti-Ballistic Missile Treaty, the 1991 Strategic Arms Reduction Treaty, the 2002 Strategic Offensive Reductions Treaty, and the 2010 New Strategic Arms Reduction Treaty.[95] Such agreements are taken to prevent destabilizing quantitative arms races wherein each country expands its arsenal with the aim of gaining a numerical advantage over the other. These agreements do, however, have the potential to increase the pressure on states to qualitatively advance their nuclear arsenals.[96] This pressure is principally caused by two main phenomena: the increased importance of individual missiles when arsenal size is limited, and the function of specific systems as bargaining chips in arms control and limitation efforts.

The first manner in which arms control, limitation, and reduction agreements can pressure states to qualitatively improve their nuclear arsenals is simple: fewer nuclear weapons means that, for a deterrent to be credible, each warhead must have a greater chance of being delivered successfully. Agreements on mutual arms reductions or shared limitations control this to an extent. They place limitations on all parties' nuclear forces, meaning that – in theory, at least for states with comparable arsenals like Russia and the United States – the odds of a state party to the agreement being able to use its nuclear weapons to completely destroy those of another, are no different than if neither state was party to the agreement.

In practice, however, factors outside the scope of these agreements do affect the ability of a warhead to reach its intended target. These factors include capabilities that may or may not be developed with the intent to affect the balance of nuclear forces. Missile defense systems, for instance, are explicitly designed to limit the odds of an adversary carrying out a successful missile strike with either a conventional or nuclear warhead. Conventional military capabilities, however, could also be seen as a threat to a state's ability to deliver a nuclear warhead. This is especially true of capabilities the United States has designated for the Prompt Global Strike mission. That mission seeks "to provide the United States with the ability to strike targets anywhere on Earth with conventional weapons in as little as an hour, without relying on forward based forces."[97] Russia has voiced concern over the potential use of Prompt Global Strike capabilities, in conjunction with missile defense assets, to cripple Russia's nuclear deterrent via a counterforce strike.[98] And, as mentioned previously in this chapter, both Russia's development of the Sarmat with its increased destructive potential (because of its MIRVed payload) and the Avangard hypersonic glide vehicle have been framed as a response to the development of such capabilities. This demonstrates how limits on nuclear delivery vehicles, in combination with external factors impacting the chances of a warhead successfully reaching its intended target, can drive states to seek qualitative advancements to their nuclear arsenals.

The second means by which arms control, limitation, and reduction agreements can incentivize the acquisition of new or improved nuclear capabilities is through their potential use as bargaining chips in advance of further agreements. This may be what the United States has in mind with its plans to acquire new non-strategic nuclear weapons. The 2018 *Nuclear Posture Review*, after all, notes that the United States may reconsider its acquisition of a new nuclear-equipped SLCM if Russia is found to return to compliance with the INF Treaty, while also expressing hope that the planned SLCM acquisition may make Russia amenable to a drawdown of its non-strategic nuclear forces.[99]

This casts the planned SLCM acquisition as grounded in the notion that parity (in this case in low yield capability rather than expressly parallel systems) is a necessary precondition for negotiated restrictions on nuclear capabilities. While this may have some merit – attributing the lack of arms control, limitation, or reduction agreements between the United States and the People's Republic of China to disparities in the size of nuclear arsenals, for instance – it also ignores several dangers associated with developing weapons systems as bargaining chips. Principle among these is that the anticipated bargain will fail to materialize and the state will retain a system that it never necessarily viewed as serving a strategic use. This may be the fate of the planned American nuclear-equipped SLCM. The United States withdrawal from the INF Treaty, announced in February 2019 by President of the United States Donald Trump,[100] will likely scuttle any hopes of using the SLCM acquisition to drive Russia back into compliance with that treaty. Indeed, the Russian reaction to the U.S. withdrawal has been to seek land-based modifications for the sea-based Kalibr launch system (a potentially nuclear-capable cruise missile)[101] and "land-based launchers for hypersonic intermediate-range and shorter-range missiles."[102] So, if the goal of seeking a low-yield SLCM and withdrawing from the INF Treaty was to discourage Russia from pursuing such systems, it appears to have failed spectacularly and may saddle the United States with a nuclear weapons system of dubious strategic worth.

The withdrawal from that treaty by the United States also casts doubt on American commitment to any other bilateral or multilateral agreement. This makes it unlikely that Russia would enter into an agreement with the United States whereby it would reduce its non-strategic nuclear forces. Instead, as explored earlier in this chapter, the expansion of the United States' non-strategic nuclear capabilities appears to have prompted Russia to re-evaluate its own – potentially increasing the danger to the United States.

The actual effect of arms control, limitation, and reduction treaties in constraining state behavior is open to interpretation. The American withdrawal from the 1972 ABM Treaty and the INF Treaty can be taken to show that states will not let international agreements prevent them from acquiring capabilities that they believe to be in their best interest. The alleged violations of the INF Treaty[103] that led the United States to withdraw from the agreement could be seen as further evidencing this point, more so in that each side is aware the other may act counter to its commitments than in the specific allegations themselves. This implies that arms treaties, at least in the case of the United States and Russia, may have less of a dampening effect on the arms spiral – the competition between countries for ever greater military capabilities and the comparative advantage those would convey that threatens to escalate to military conflict – than has been previously thought.

Future prospects

Despite the stated commitment by both the United States and Russia to a world free of nuclear armaments, each country's efforts to modernize and qualitatively improve those armaments mean that they will remain a key aspect of geopolitics for years to come. Moreover, the effects of arms control, reduction, or limitation measures – as examined in this chapter – don't necessarily de-incentivize the United States or Russia from attempting to gain a technological advantage over the other. Instead, they can prompt unnecessary competition in the development of nuclear armaments. For this reason, arms control, reduction, and limitation measures are not in and of themselves the most promising avenue for reducing pressures pushing toward what former United States Ambassador to Russia Michael McFaul termed a "new qualitative arms race."[104]

Fortunately, there are steps that the United States and Russia can take to decrease the urgency of competitive qualitative improvements to their nuclear arsenals. Both countries will undoubtedly continue making qualitative improvements to their nuclear arsenals. But these improvements should be limited to keeping pace with the technological imperatives necessitating the modernization of each country's nuclear arsenal, else they will become little more than an expensive means of ratcheting up tensions between Russia and the United States. Crucially, there are steps in this direction that each country can take without jeopardizing their security.

Russia, for starters, could clarify the roles of its dual-capable systems, i.e. whether Russia actually has a role assigned to them in its nuclear doctrine. Confirming a nuclear role for these systems would possibly elicit further reaction from the United States but, as detailed earlier in this chapter, the United States is already reacting to a perceived nuclear threat from these systems. If Russia does not currently rely on these systems as a fixture of its nuclear strategy, publicly stating as much could prompt the United States to reevaluate its actions to counter what it views as an 'escalate to de-escalate' strategy. Russia, if it chooses to publicly deny a nuclear role for its dual-capable systems, could take steps to make those systems incompatible with nuclear warheads. This would serve to enhance the credibility of Russia's denials of the existence of an 'escalate to de-escalate' strategy.

The United States, for its part, could abandon its plans to acquire new non-strategic nuclear weapons. As detailed earlier in this chapter, the acquisition of a new nuclear submarine-launched cruise missile does not provide a functional benefit to the American nuclear deterrent. Parity for the parity's sake is expensive, utterly unnecessary, and would increase the profile of nuclear weapons, thereby prompting countries to invest ever more in those weapons. For the same reason, the United States should not follow its withdrawal from the INF Treaty with the development of an intermediate-range ballistic missile. As government officials have pointed out, such a system would be redundant – existing capabilities are already performing the roles that would be assigned to it.[105] Its only relatively unique function would be as leverage for mutual future reductions of similar systems. As detailed in the prior section, this too would likely drive Russia to further development of military capabilities that could increase its threat to the United States.

Finally, both Russia and the United States could dial down their rhetoric surrounding nuclear weapons. Both the United States and Russia have a history of speaking of their nuclear capabilities in hyperbole. To take two relatively recent examples, President Trump has oversold the efficacy of missile defense systems – claiming a 97% success rate, no less[106] – and President Putin deemed Russia's Kh-101/102 ALCM "invincible against all existing and prospective missile defence and counter-air defence systems."[107] Either country's use of such language may serve to exaggerate the threat it poses to the other, convincing them to invest in additional capabilities to regain parity when parity was never lost in the first place. This would create an escalatory dynamic with destabilizing potential, pushing the United States and Russia deeper into an arms spiral.

The intended effect of such steps would not be to move toward a vision of a world free of nuclear weapons. Nor would it be to leave Russia or the United States armed, in the unlikely event of a nuclear conflict, with obsolete weapons. Rather, the intent should be to constrain further investment in nuclear capabilities to what is necessary to maintain a credible deterrent.

Notes

1 Barry Buzan and Eric Herring, *The Arms Dynamic in World Politics* (Boulder, CO: Lynne Rienner Publishers, Inc., 1998), 83.

2 Ibid.

3 See Pavel Podvig, "The Truth About Russia's Military 'Resurgence'," *Bulletin of the Atomic Scientists*, January 28, 2018, https://thebulletin.org/2008/01/the-truth-about-russias-military-resurgence/.

Alternatively, see Dmitry Gorenburg, Alexander Savelyev, and Alexander Stukalin, "Medvedev's Defense and Security Policy," *Valdai Discussion Club*, May 10, 2012, http://valdaiclub.com/a/highlights/medvedev_s_defense_and_security_policy/?sphrase_id=430834.

4 Current movements by the U.S. Congress to restrict deployment of low-yield nuclear weapons, for example. See Rachel S. Cohen, "HASC Fiercely Debates Low-Yield Nuclear Weapons at Markup," *Air Force Magazine*, June 12, 2019, www.airforcemag.com/Features/Pages/2019/June%202019/HASC-Fiercely-Debates-Low-Yield-Nuclear-Weapons-at-Markup.aspx.

See also Ben Werner and John Grady, "Funding for Low-Yield Nukes Could Be Sticking Point in Budget Bill Debate," *USNI News*, June 11, 2019, https://news.usni.org/2019/06/11/funding-for-low-yield-nukes-could-be-sticking-point-in-budget-bill-debate.

5 Buzan and Herring, *The Arms Dynamic in World Politics*, 86.
6 Ibid., 82.
7 "Указ Президента Российской Федерации от 31.12.2015 г. № 683: О Стратегии национальной безопасности Российской Федерации [Order of the President of the Russian Federation from 31/12/2015 No. 683: About the National Defense Strategy of the Russian Federation]," President of Russia, December 31, 2015, http://kremlin.ru/acts/bank/40391/.

Embassy of the Russian Federation to the United Kingdom of Great Britain and Northern Ireland, "The Military Doctrine of the Russian Federation" (Press Release), June 29, 2015, https://rusemb.org.uk/press/2029

"Военная доктрина Российской Федерации [Military Doctrine of the Russian Federation]," President of Russia, February 5, 2010, http://kremlin.ru/supplement/461.

"Указ Президента Российской Федерации от 21.04.2000 г. № 706: Об утверждении Военной доктрины Российской Федерации [Order of the President of the Russian Federation from 21/04/2000 No. 706: On Approval of the Military Doctrine of the Russian Federation]," President of Russia, April 21, 2000, http://kremlin.ru/acts/bank/15386.

"The Basic Provisions of the Military Doctrine of the Russian Federation," *Federation of American Scientists*, accessed June 11, 2019, https://fas.org/nuke/guide/russia/doctrine/russia-mil-doc.html

8 "National Security Concept of the Russian Federation," *Ministry of Foreign Affairs of the Russian Federation*, January 10, 2000, www.mid.ru/en/foreign_policy/official_documents/-/asset_publisher/CptICkB6BZ29/content/id/589768.
9 "SIPRI Military Expenditure Database: Data for All Countries from 1988–2018 in Constant (2017) USD," *Stockholm International Peace Research Institute*, accessed June 2019, www.sipri.org/databases/milex.
10 "Russian Strategic Nuclear Forces, End of 1998," *Bulletin of the Atomic Scientists* 55, no. 2 (1999), 62–63.
11 "Гибель атомной подводной лодки 'Курск'. Справка [The Death of the Nuclear Submarine 'Kursk'. Reference]," *RIA Novosti*, August 12, 2010, <https://ria.ru/spravka/20100812/264217073.html>.
12 "Russian Strategic Nuclear Forces, End of 1998."
13 Ibid., 62.
14 Note that Russia also alleges that components of American ballistic missile defense systems in Europe violate the INF Treaty, as will be discussed later in the chapter.
15 "Excerpts of Classified *Nuclear Posture Review* /S submitted to Congress on 31 December 2001," U.S. Department of Defense, January 8, 2002, 5 (from p. 25 of *Nuclear Posture Review* Report), https://fas.org/wp-content/uploads/media/Excerpts-of-Classified-Nuclear-Posture-Review.pdf.
16 "Speech and the Following Discussion at the Munich Conference on Security Policy" (Transcript), President of Russia, February 10, 2007, http://en.kremlin.ru/events/president/transcripts/24034.
17 For example: then-U.S. President Barack Obama and then-Russian President Dmitry Medvedev identified missile defense as a friction point at a bilateral meeting in 2012.

"Meeting with President of the United States Barack Obama" (Transcript), President of Russia, March 26, 2012, http://en.kremlin.ru/events/president/news/14840.

18 See, for example, comments by Russian President Vladimir Putin's 2014 and 2016.

"News Conference of Vladimir Putin," President of Russia, December 18, 2014, http://en.kremlin.ru/events/president/news/47250.

"Vladimir Putin's Annual News Conference," President of Russia, December 23, 2016, http://en.kremlin.ru/events/president/news/53573.

19 Stephen Blank, "Threats to and from Russia: An Assessment," *Journal of Slavic Military Studies* 21, no. 3 (200), 518.
20 "A Statement Regarding the Decision of the Administration of the United States to Withdraw from the Antiballistic Missile Treaty of 1972" (Transcript), President of Russia, December 13, 2001, http://en.kremlin.ru/events/president/transcripts/21444.
21 See Pavel Podvig, "The Myth of Strategic Stability," *Bulletin of the Atomic Scientists*, October 31, 2012, https://thebulletin.org/2012/10/the-myth-of-strategic-stability/.
22 Hans M. Kristensen and Matt Korda, "Russian Nuclear Forces, 2019," *Bulletin of the Atomic Scientists*, 75, no. 2 (2019), 74.
23 Missile Defense Project, "SS-27 'Sickle B' (RT-2PM2 Topol-M)," *Missile Threat*, Center for Strategic and International Studies, August 10, 2016, last modified June 15, 2018, https://missilethreat.csis.org/missile/ss-27/.
24 Hans M. Kristensen and Robert S. Norris, "Russian Nuclear Forces, 2018," *Bulletin of the Atomic Scientists* 74, no. 3 (2018), 189.
25 Luke Harding, "Russian Missile Test Adds to Arms Race Fears," *The Guardian*, May 29, 2007, www.theguardian.com/world/2007/may/30/usa.topstories3.
26 See Missile Defense Project, "SS-X-30 'Satan II' (RS-28 Sarmat)," *Missile Threat*, Center for Strategic and International Studies, May 17, 2017, last modified January 3, 2019, https://missilethreat.csis.org/missile/rs-28-sarmat/. See also Lt. Gen. Robert P. Ashley Jr., "Russian and Chinese Nuclear Modernization Trends: Remarks at the Hudson Institute" (Transcript), Defense Intelligence Agency, May 29, 2019, www.dia.mil/News/Speeches-and-Testimonies/Article-View/Article/1859890/russian-and-chinese-nuclear-modernization-trends/.
27 "First Image of RS-28 Sarmat, Russia's Largest-Ever ICBM, Unveiled (PHOTO)," *Sputnik*, October 24, 2016, https://sputniknews.com/military/201610241046655887-sarmat-image-declassified/.
28 Kristensen and Korda, "Russian Nuclear Forces, 2019," 78.
29 Robert S. Norris and Hans M. Kristensen, "Russian Nuclear Forces, 2003," *Bulletin of the Atomic Scientists* 59, no. 4 (2003), 72.
30 Missile Defense Project, "SS-M-32 Bulava," *Missile Threat*, Center for Strategic and International Studies, August 10, 2016, last modified June 15, 2018, https://missilethreat.csis.org/missile/ss-n-32-bulava/.
31 Franz-Stefan Gady, "Russia's Navy to Receive Upgraded Variant of New Ballistic Missile Sub in 2019," *The Diplomat*, May 9, 2018, https://thediplomat.com/2018/05/russias-navy-to-receive-upgraded-variant-of-new-ballistic-missile-sub-in-2019/.
32 "Источник: Россия построит еще шесть атомных стратегических подлодок класса 'Борей-А' [Source: Russia Will Build Another Six Nuclear Strategic Submarines of the Class 'Borei-A']," *TASS*, May 21, 2018, https://tass.ru/armiya-i-opk/5218417.
33 Kristensen and Norris, "Russian Nuclear Forces, 2018," 189.
34 Calculated as 'number SLBM warheads' divided by 'total strategic offensive warheads' and based on a review of the 2000–2019 Russian nuclear forces entries in the 'Nuclear Notebook' published annually in the *Bulletin of the Atomic Scientists*.
35 Calculated in the same manner as preceding endnote.
36 Kristensen and Norris, "Russian Nuclear Forces, 2019," 79.
37 Andrei Akulov, "Russian Kh-101 Air-to-Surface Cruise Missile: Unique and Formidable," Strategic Culture Foundation, October 19, 2016, www.strategic-culture.org/news/2016/10/19/russian-kh-101-air-to-surface-cruise-missile-unique-and-formidable.html.
38 Missile Defense Project, "Kh-101/Kh-102," *Missile Threat*, Center for Strategic and International Studies, October 26, 2017, last modified June 15, 2018, https://missilethreat.csis.org/missile/kh-101-kh-102/.
39 "Russia's Advanced Strategic Bomber May Perform Debut Flight in Late 2018," *TASS*, August 4, 2016, www.tass.ru/en/defense/892376.

40 "World's Most Famous Stealth Aircraft," *TASS*, March 31, 2016, https://tass.ru/en/defense/866381.
41 Hans M. Kristensen, "Despite Obfuscations, New START Data Shows Continued Value of Treaty," *Federation of American Scientists*, April 10, 2019, https://fas.org/blogs/security/2019/04/newstartdata-2019-1/.
42 "Presidential Address to the Federal Assembly" (Transcript), President of Russia, March 1, 2018, http://en.kremlin.ru/events/president/transcripts/messages/56957.
43 "Presidential Address to Federal Assembly" (Transcript), President of Russia, February 20, 2019, http://en.kremlin.ru/events/president/transcripts/messages/59863.
44 Sean Gallagher, "Russia Launches Sub That Will Carry Doomsday Nuke Drone Torpedo," *Ars Technica*, April 25, 2019, https://arstechnica.com/information-technology/2019/04/russia-launches-sub-that-will-carry-doomsday-nuke-drone-torpedo/.
45 "Russian Navy to Put Over 30 Poseidon Strategic Underwater Drones on Combat Duty – Source," *TASS*, January 12, 2019, http://tass.com/defense/1039603.
46 Geoff Brumfiel, "Russia's Nuclear Cruise Missile Is Struggling to Take Off, Imagery Suggests," *NPR*, September 25, 2018, www.npr.org/2018/09/25/649646815/russias-nuclear-cruise-missile-is-struggling-to-take-off-imagery-suggests.
47 "Avangard Hypersonic Missiles Replace Rubezh ICBMs in Russia's Armament Plan Through 2027," *TASS*, March 22, 2018, http://tass.com/defense/995628.
48 Julian Cooper, "The Funding of Nuclear Weapons in the Russian Federation," October 2018, 9, www.ccw.ox.ac.uk/blog/2018/10/1/the-funding-of-nuclear-weapons-in-the-russian-federation-by-julian-cooper.
49 "Nuclear Posture Review Report," United States Department of Defense, April 2010, 23, https://dod.defense.gov/Portals/1/features/defenseReviews/NPR/2010_Nuclear_Posture_Review_Report.pdf.
50 Ibid., 27.
51 Ibid., 24.
52 Ibid., 30.
53 Ibid.
54 United States Department of Defense, Office of the Secretary of Defense, "Nuclear Posture Review 2018," February 2018, 45–46, https://media.defense.gov/2018/Feb/02/2001872886/-1/-1/1/2018-NUCLEAR-POSTURE-REVIEW-FINAL-REPORT.PDF.
55 Ibid., 47.
56 Ibid., 31.
57 Amy F. Woolf, "Nonstrategic Nuclear Weapons," *Congressional Research Service*, February 13, 2018, 10.
58 Ibid., 11–12.
59 Hans M. Kristensen, "I. U.S. Nuclear Forces," in Stockholm International Peace Research Institute (ed.), *SIPRI Yearbook 2018* (Oxford, United Kingdom: Oxford University Press, 2018), 238, www.sipri.org/sites/default/files/SIPRIYB18c06.pdf.
60 "Remarks by Mikhail Ulyanov, Director of the Foreign Ministry Department for Non-Proliferation and Arms Control and Representative of the Russian Federation at the First Committee of the 71st Session of the UNGA, Within the General Debate, New York, October 4, 2016" (Transcript), Ministry of Foreign Affairs of the Russian Federation, October 5, 2016, www.mid.ru/en/web/guest/general_assembly/-/asset_publisher/lrzZMhfoyRUj/content/id/2488843.
61 Kristensen, "I. U.S. Nuclear Forces," 238–239.
62 U.S. Department of Defense, Office of the Secretary of Defense, "Nuclear Posture Review 2018," 54.
63 Ibid., 55.
64 Ibid., 48.
65 Ibid., 50.

66 See Cohen, "HASC Fiercely Debates Low-Yield Nuclear Weapons at Markup." See also Werner and Grady, "Funding for Low-Yield Nukes Could be Sticking Point in Budget Bill Debate."
67 U.S. Department of Defense, Office of the Secretary of Defense, "Nuclear Posture Review 2018," *op. cit.*, 53.
68 Ibid.
69 Missile Defense Project, "SSC-8 (Novator 9M729)," *Missile Threat*, Center for Strategic and International Studies, October 23, 2018, last modified January 23, 2019, https://missilethreat.csis.org/missile/ssc-8-novator-9m729/.
70 Tytti Erästö and Petr Topychkanov, "Russian and U.S. Policies on the INF Treaty Endanger Arms Control," *Stockholm International Peace Research Institute*, June 15, 2018, www.sipri.org/commentary/topical-backgrounder/2018/russian-and-us-policies-inf-treaty-endanger-arms-control.
71 Bruno Tetrais, "Russia's Nuclear Policy: Worrying for the Wrong Reasons," *Survival* 60, no. 2 (2018), 36.
72 U.S. Department of Defense, Office of the Secretary of Defense, "Nuclear Posture Review 2018," 9.
73 See Robert S. Norris and Hans M. Kristensen, "Russian Nuclear Forces, 2010," *Bulletin of the Atomic Scientists* 66, no. 1 (2010), 74–81. See also Kristensen and Norris, "Russian Nuclear Forces, 2018."
74 "Российские военные применяли в Сирии комплекс 'Искандер' [Russian Military Used Iskander Complex in Syria]," *RIA Novosti*, December 22, 2017, https://ria.ru/syria/20171222/1511520986.html.
75 Olga Oliker and Andrey Baklitskiy, "The Nuclear Posture Review and Russian 'De-Escalation': A Dangerous Solution to a Nonexistent Problem," *War on the Rocks*, February 20, 2018, https://warontherocks.com/2018/02/nuclear-posture-review-russian-de-escalation-dangerous-solution-nonexistent-problem/.
76 Alternatively referred to as 'escalate to win'. See: General John E. Hyten, "Military Reporters and Editors Association Conference – Keynote Speech" (Transcript) U.S. Strategic Command, March 31, 2017, www.stratcom.mil/Media/Speeches/Article/1153029/military-reporters-and-editors-association-conference-keynote-speech/.
77 Olga Oliker, "Russia's Nuclear Doctrine: What We Know, What We Don't, and What That Means," *Center for Strategic & International Studies*, May 5, 2016, 2, https://csis-prod.s3.amazonaws.com/s3fs-public/publication/160504_Oliker_RussiasNuclearDoctrine_Web.pdf.
78 Hans M. Kristensen and Robert S. Norris, "Russian Nuclear Forces, 2017," *Bulletin of the Atomic Scientists* 73, no. 2 (2017), 115–126.
79 See Dmitry Adamsky, "Nuclear Incoherence: Deterrence Theory and Non-Strategic Nuclear Weapons in Russia," *Journal of Strategic Studies* 37, no. 1 (2014), 91–134.
    Alternatively, see Pavel Podvig, "What to Do About Nuclear Weapons," *Bulletin of the Atomic Scientists*, February 25, 2010, https://thebulletin.org/2010/02/what-to-do-about-tactical-nuclear-weapons/.
80 U.S. Department of Defense, Office of the Secretary of Defense, "Nuclear Posture Review 2018," 8.
81 See: Nikolai N. Sokov, "Why Russia Calls a Limited Nuclear Strike 'De-Escalation'," *Bulletin of the Atomic Scientists*, March 13, 2014, https://thebulletin.org/2014/03/why-russia-calls-a-limited-nuclear-strike-de-escalation/.
82 See: Major General V.I. Levshin, Colonel A.N. Nedelin, and Colonel M.E. Sosnovsky, "О применении ядерного оружия для деэскалации военных действий [On the Use of Nuclear Weapons for the De-Escalation of Hostilities]," *Voennaya Mysl* 5–6, no. 3 (1999), 34–37, http://militaryarticle.ru/zarubezhnoe-voennoe-obozrenie/1999-zvo/8995-o-primenenii-jadernogo-oruzhija-dlja-dejeskalacii.

83 On this point, the 2014 *Quadrennial Defense Review* uses similar language to that later used by then Secretary of Defense Ashton Carter in 2016. It asserts that unspecified potential nuclear-armed adversaries may be under the belief that they can use nuclear weapons in the event of a failed conventional attack on American or allied forces.

"Quadrennial Defense Review 2014," U.S. Department of Defense, March 4, 2014, https://archive.defense.gov/pubs/2014_Quadrennial_Defense_Review.pdf.

84 "Nuclear Posture Review Report," 28.
85 U.S. Department of Defense, Office of the Secretary of Defense, "Nuclear Posture Review 2018," 54.
86 Oliker, "Russia's Nuclear Doctrine," 12.
87 U.S. Department of Defense, Office of the Secretary of Defense, "Nuclear Posture Review 2018," 54.
88 Ash Carter, "Remarks by Secretary Carter to Troops at Minot Air Force Base, North Dakota" (Transcript), U.S. Department of Defense, September 26, 2016, https://dod.defense.gov/News/Transcripts/Transcript-View/Article/956079/remarks-by-secretary-carter-to-troops-at-minot-air-force-base-north-dakota/.
89 "Comment by the Information and Press Department on the U.S. Defence Secretary's Remarks on Nuclear Deterrence" (Press release), Ministry of Foreign Affairs of the Russian Federation, September 29, 2016, www.mid.ru/en/foreign_policy/news/-/asset_publisher/cKNonkJE02Bw/content/id/2478696.
90 "Comment by the Information and Press Department on the New U.S. Nuclear Posture Review" (Press release), Ministry of Foreign Affairs of the Russian Federation, February 3, 2018, www.mid.ru/en/web/guest/kommentarii_predstavitelya/-/asset_publisher/MCZ7HQuMdqBY/content/id/3054726.
91 Hans M. Kristensen, "II. Russian Nuclear Forces," in Stockholm International Peace Research Institute (ed.), *SIPRI Yearbook 2018* (Oxford: Oxford University Press, 2018), 249, www.sipri.org/sites/default/files/SIPRIYB18c06.pdf.
92 Ibid.
93 "Remarks by Foreign Minister Sergey Lavrov at the UN Security Council Meeting, September 26, 2018" (Transcript), Embassy of the Russian Federation to the United Kingdom of Great Britain and Northern Ireland, September 27, 2018, www.rusemb.org.uk/article/526.
94 Nicholas Graham, "Obama Prague Speech on Nuclear Weapons," *HuffPost*, May 6, 2009, last modified May 25, 2011, www.huffpost.com/entry/obama-prague-speech-on-nu_n_183219.
95 "U.S.-Russian Nuclear Arms Control Agreements at a Glance," *Arms Control Association*, updated February 2019, www.armscontrol.org/factsheets/USRussiaNuclearAgreements.
96 Buzan and Herring, *The Arms Dynamic in World Politics*, 225. Buzan and Herring note something similar in *The Arms Dynamic in World Politics* when they observe that, during the Cold War, Soviet possession of a capability strengthened domestic arguments for why the United States needed such a capability (or a better version of that capability) and vice versa. Specifically, they point to arguments for chemical and anti-satellite weapons in the United States and cruise missiles and enhanced-radiation warheads in the Soviet Union.
97 Amy F. Woolf, "Conventional Prompt Global Strike and Long-Range Ballistic Missiles: Background and Issues," *Congressional Research Service*, April 6, 2018, 1.
98 "Presidential Address to the Federal Assembly," President of Russia, December 12, 2013, http://en.kremlin.ru/events/president/transcripts/messages/19825.
99 U.S. Department of Defense, Office of the Secretary of Defense, "Nuclear Posture Review 2018," 55.

100 "Statement from the President Regarding the Intermediate-Range Nuclear Forces (INF) Treaty" (Press Release), The White House, February 1, 2019, www.whitehouse.gov/briefings-statements/statement-president-regarding-intermediate-range-nuclear-forces-inf-treaty/.
101 Missile Defense Project, "SS-N-30A (3M-14 Kalibr)," *Missile Threat*, Center for Strategic and International Studies, August 11, 2016, last modified June 15, 2018, https://missilethreat.csis.org/missile/ss-n-30a/.
102 "Meeting with Sergei Lavrov and Sergei Shoigu," President of Russia, February 2, 2019, http://en.kremlin.ru/events/president/transcripts/statements/59763.
103 See "Director of National Intelligence Daniel Coats on Russia's INF Treaty Violation" (Transcript), Office of the Director of National Intelligence, November 30, 2018, www.dni.gov/index.php/newsroom/speeches-interviews/item/1923-director-of-national-intelligence-daniel-coats-on-russia-s-inf-treaty-violation.
104 Michael McFaul, "Russia as It Is," *Foreign Affairs*, 97, no. 4 (2018), 82–91, www.foreignaffairs.com/articles/russia-fsu/2018-06-14/russia-it.
105 Daryl G. Kimball and Kingston Reif, "Trump's Counterproductive Decision to 'Terminate' the INF Treaty," *Arms Control Association Issue Briefs* 10, no. 9 (October 21, 2018), www.armscontrol.org/issue-briefs/2018-10/trumps-counterproductive-decision-terminate-inf-treaty.
106 Fred Kaplan, "The Worst Defense," *Slate*, October 17, 2017, https://slate.com/news-and-politics/2017/10/trump-claims-that-u-s-missile-defense-works-97-percent-of-the-time-it-doesnt.html.
107 "Presidential Address to the Federal Assembly" (Transcript), President of Russia, March 1, 2018, http://en.kremlin.ru/events/president/news/56957.

5 Chinese nuclear strategy

Susan Turner Haynes

A recent RAND report characterized the study of Chinese nuclear weapons as a "boutique" endeavor in international relations, due, in part, to the relatively small size of China's nuclear force.[1] Indeed, the nearly 300 nuclear warheads in China's arsenal pale in comparison to the thousands held by the United States and Russia, but this discrepancy belies the steady increase and modernization of China's nuclear weapons. Over the past two decades, the number of Chinese nuclear warheads has more than doubled, and scholarship on the subject has also surged. It is thus no longer appropriate to describe China as the "forgotten nuclear power," nor is it necessarily accurate to say its study remains relegated to a niche group of academics.[2] As China has increased the size, sophistication, and salience of its nuclear force, more people have begun to pay attention – including U.S. policymakers. This is evidenced most recently by the 2018 U.S. *Nuclear Posture Review*, which, for the first time, describes China as a "potential adversary" of the United States. The report makes clear that the U.S. can no longer assume Chinese benevolence due to China's ongoing pursuit of new nuclear capabilities, its contravention of international norms, and its display of aggressive behavior in other domains. Moreover, since China has offered little to no insight into the intent of its nuclear expansion and modernization, the U.S. has been left to conclude that China is pursuing a more aggressive nuclear strategy.

This chapter examines the available open-source evidence supporting this interpretation of Chinese intent. The conclusion reached upon the review of primary sources is that China's nuclear strategy has evolved in tandem with the gradual growth and diversification of its nuclear arsenal. More specifically, evidence indicates that China seeks to transition from a purely defensive nuclear posture with relatively minimal options for retaliatory responses, to a nuclear posture that is more flexible and able to engage in nuclear war-fighting. In the parlance of the field, this is referred to as "minimum deterrence" and "limited deterrence," respectively. Since it is necessary to understand these terms and their implications prior to exploring evidence in support of a transition, this chapter begins by clarifying these concepts and contextualizing the conversation amidst the contemporary academic debate. The subsequent sections review the primary source material in support of China's strategic transition. The chapter ends with an assessment of what the future might hold for China as a nuclear power.

The academic debate

For much of China's nuclear history, its policymakers eschewed using the word "deterrence" to describe the state's nuclear strategy. Scholars in the West, however, were less reserved in doing so, and a debate emerged as to what *type* of deterrence China chose to employ. This debate was first outlined by Alastair Iain Johnston in 1995.[3] In particular, the question he explored was whether China subscribed to a strategy of "minimum deterrence" or whether it was more likely to employ a strategy of "limited deterrence." Though scholars in the ensuing years have added nuance to the conversation, the general contours of the debate remain the same.

The first school of thought contends that China's relatively small nuclear structure reflects its commitment to use nuclear weapons only for self-defense, deterring adversarial aggression by threatening to strike back with whatever weapons remain after a first strike. Scholars in this camp argue that the current modernization and growth of China's nuclear force indicate only a recalibration of what China considers to be necessary for minimum deterrence in light of the capabilities of potential adversaries. Minimum force levels, they say, are relative; though China's force structure has shifted, its strategy remains consistent.[4]

The second school of thought argues that China's changing security environment and its increased number of external threats actively shape its nuclear force structure *and* its strategy. Scholars on this side of the debate argue that China is not satisfied with relying upon a small, survivable nuclear deterrent, but wants to build a force capable of acting in the event that deterrence fails. Under limited deterrence, China would have more at its disposal than a limited retaliatory strike, and it would have the option to engage in nuclear war if necessary. The primary goal of both strategies is to deter adversarial aggression, but the means by which they do so differ. They also differ in several other key aspects, as outlined below.

Minimum deterrence

In his original explanation, Johnston describes minimum deterrence as a strategy where states aim to deter a foreign nuclear attack by threatening nuclear retaliation. This threat, to be credible, does not require a state to match or exceed its adversary's nuclear capabilities, but only that it has enough nuclear weapons to deliver a devastating counterattack. Even a handful of weapons, it is assumed, can have this effect. A logical corollary is for the state to communicate that it intends only to use nuclear weapons in response to a nuclear attack and not in retaliation for non-nuclear aggression. This is critical to the state's survival, because launching a first strike invites a counterattack by an adversary's remaining weapons. A state that is numerically disadvantaged thus has no incentive to strike first – even if it is against an enemy's nuclear forces – since it knows even a full-scale first strike will leave many more weapons for a counterstrike. In fact, a state in this position has a very strong *disincentive* to consider this option, since a first strike would be akin to kicking a hornet's nest, and may be a suicide mission.

From the beginning, Chinese statements indicate that the state did not intend to be a nuclear provocateur. Instead, China would use its weapons as instruments of "peace," fending off nuclear war with the two superpowers. A communique issued immediately after China's first nuclear test makes this clear. "On the question of nuclear weapons," it said, "China will commit neither the error of adventurism, nor the error of capitulation."[5] Absent a state codification of Chinese nuclear strategy, most Chinese scholars use these words and the words of Chinese officials to support the claim that China's nuclear thinking has long been characterized by restraint.[6] Many Western scholars have also argued this position. In their seminal work on China's nuclear force, for instance, John Wilson Lewis and Xue Litai, use the statements of Mao in the 1960s and 1970s to highlight "seven principles" guiding China's nuclear force, including limiting the use of nuclear weapons to a second strike, abstaining from the development of tactical weapons, focusing on weapon quality over quantity, maintaining force diversity, prioritizing civilian targets, and preparing for quick recovery following a nuclear war.[7] Jeffrey Lewis, writing nearly two decades later, provides a similar characterization of China's nuclear strategy based upon the memoirs of Nie Rongzhen and the biographical essays of others involved in China's nuclear, missile, and space program.[8] More recently, Renny Babiarz uses historical Chinese military publications to suggest that the strategic ambitions of China's nuclear force are mitigated by the domestic strategic culture developed under Mao.[9]

Other scholars arguing for minimum deterrence cite more recent evidence, such as the Chinese White Papers. These papers, published biennially beginning in 1998, present a reverberating chorus praising China's decision to maintain a "lean and effective" nuclear force. They also repeatedly emphasize China's decision not to use nuclear weapons first, not to enter into an arms race, not to use nuclear weapons in the arena of war, and not to deploy forces outside of Chinese territory.[10] Alongside such contestations, there are words advocating for the state's defensive military orientation, delayed nuclear response, and the elimination of nuclear weapons world-wide. The state's unchanging rhetoric has convinced those like Chinese military scholar Yao Yunzhu that China's nuclear strategy has always been one of minimum deterrence. In fact, Yao claims "of all the nuclear states, the nuclear policy of China has so far been the most consistent."[11]

Many scholars would characterize China's force structure in the same terms, claiming that China's arsenal has remained among the most consistent in terms of relative size. While China's force has grown, it has never rivaled that of the United States and Russia. This, they say, is evidence that China takes a different strategic approach. If a state employs a strategy of minimum deterrence, then the emphasis is less on a quantitative force threshold than it is on necessary qualitative force characteristics. In particular, effective deterrence requires a state to have nuclear weapons that are survivable, deliverable, and able to penetrate enemy defenses. Scholars in this camp perceive China's recent nuclear force decisions as logical consequences of a state trying to satisfy the criteria of minimum deterrence amidst a shifting security landscape.[12] They also point out that China lacks some of the necessary capabilities for a more advanced nuclear strategy, such as short-range nuclear weapons ideal for battlefield use and missiles mated with warheads ready for quick launch.[13]

Limited deterrence

Some scholars, like Alastair Johnston, believe minimum deterrence never accurately characterized Chinese thinking on nuclear weapons, because it goes against the *realpolitik* culture underpinning the Chinese state.[14] China, like other states, makes its nuclear force decisions contingent upon the relative nuclear capabilities of other states and its external security environment. When these things shift, says Johnston, we should expect China's nuclear strategy and nuclear force structure to do the same. Moreover, Johnston argues that Chinese leaders have always valued the military utility of nuclear weapons and engaged in thought exercises regarding the role of nuclear weapons in war.[15] According to Johnston, this role was initially minimal not because of China's strategic restraint, but because of its *technological* constraints.

Johnston claims that China's strategic ambition has long been to employ a strategy of "limited deterrence," a strategy which "requires sufficient counterforce and countervalue tactical, theater, and strategic nuclear forces to deter escalation of conventional or nuclear war."[16] This statement highlights several distinctive features of limited deterrence, including the expanded aims of the strategy (i.e., deterring both nuclear *and* conventional attacks) and the expanded means of the strategy (i.e., targeting both countervalue *and* counterforce targets). This means that a state is open to using nuclear weapons to deter the escalation of military conflict by threatening to strike both civilian and military targets in and outside of enemy territory. States could use their long-range missiles to strike a portion of an adversary's population just as they could use more weapons of the same type to strike an adversary's nuclear force or command centers. States could also consider using short-range missiles in a battlefield scenario against an adversary's military forces.

The largest caveat of Johnston's claim was what he called the "doctrine-capabilities gap" – meaning that China did not have the nuclear capabilities at the time to satisfy its desire for a limited deterrence strategy.[17] For limited deterrence to be an operational strategy, Johnston claimed China needed: "a greater number of tactical, theater, and strategic nuclear weapons that are accurate enough to hit counterforce targets, are mobile, can be used in the earliest stages of a nuclear crisis, and . . . are capable of penetrating ballistic missile defense systems."[18] At the time of Johnston's writing, China did not have such capabilities. Today, China is much closer. In fact, this is often the primary reason some scholars have come to doubt China's claim of minimum deterrence.[19]

Over the past twenty years, China has pursued significant upgrades to its nuclear arsenal, including shifting from liquid- to solid-fueled missiles, adding new intercontinental ballistic missiles to its force, loading older missiles with more warheads, and enhancing its sea-based nuclear deterrent with a new submarine-launched ballistic missile and carrier fleet. While some analysts see these advances as an attempt by China to increase its survivability and ability to penetrate enemy defenses pursuant to the necessary conditions of minimum deterrence, others argue that these technological developments signal a broader strategic shift.[20]

Another factor emphasized by scholars is the location of China's nuclear-capable Intermediate-Range Ballistic Missiles (IRBMs). According to Michael Mazza and Dan Blumenthal, China's Dongfeng-21 IRBMs are believed to be based in areas within striking distance of India, South Korea, and Japan.[21] Since the latter two are not nuclear weapon states, Mazza and Blumenthal surmise that the purpose of these locations could be to target U.S. military bases in these countries. The latest DOD Report on China's Military Strength includes a similar assessment.[22] It also mentions the risk these weapons pose to naval forces in the western Pacific and Indian Oceans and the South China Sea.[23] China's latest IRBM, the Dongfeng-26, extends this threat to U.S. bases in Guam.[24] Together, the Dongfeng-21 and Dongfeng-26 represent a formidable counterweight to U.S. power projection in the region, and, also have been considered by some to signal China's strategic shift to counterforce targeting.[25]

An additional element arousing scholarly suspicion is China's nuclear use policy. Immediately after China tested its first atomic bomb in 1964, Chairman Mao announced that China would use these weapons only in retaliation for a nuclear strike against the Chinese homeland. This policy, known as No First Use (NFU), was unique among the nuclear weapon states at the time, and it remains in place today. Despite Chinese officials' consistent reaffirmation of the policy, however, some scholars find evidence that China's commitment is waning. James Acton, for instance, looks to the most authoritative of China's publications on the subject and finds China's NFU policy conspicuously absent from the 2012 Defense White Paper.[26] Mazza and Blumenthal look to less official channels, highlighting an appreciable difference in the level of commitment to NFU expressed by China's military leaders and writings.[27] Others, like Liping Xia, are more ambiguous on this point, but mention that China's NFU policy is contested in some circles in China.[28]

In assessing China's nuclear strategy, here as elsewhere, words are paramount. Scholars on both sides of the debate take great care to analyze the words spoken by Chinese representatives and the words published by the Chinese press; yet ironically the word choice of the scholars themselves has recently come under scrutiny. Some claim, for instance, that framing the debate in terms of "minimum deterrence" and "limited deterrence" falsely characterizes the situation. Instead, they argue the truth is somewhere in between.

Other perspectives

Several scholars argue that China has shifted in many areas away from minimum deterrence, but cannot be accurately described as employing or even necessarily aspiring to a strategy of limited deterrence. Michael Swain, for instance, acknowledges that there has been "gradual movement" in China away from minimum deterrence, but that China retains what he calls a "nonwarfighting" strategy.[29] Similarly, James C. Mulvenon admits that Chinese nuclear strategy can no longer accurately be characterized as "minimal deterrence," but that it has evolved into what he calls "credible minimal deterrence," a strategy which also incorporates counterforce targeting.[30] Gill, Mulvenon, and Stokes provide another perspective by abandoning the continuum framework altogether and instead arguing that modern-day China actually employs three *separate* nuclear doctrines.[31] They claim credible deterrence best describes the doctrine China employs for the continental U.S. and Russia; limited deterrence best suits China's theater nuclear forces; and a doctrine of "offensively configured, preemptive, counterforce warfighting" applies to China's conventional missiles. Taylor Fravel and Evan Medieros's argument provides similar nuance to the conversation, as they introduce the term "assured retaliation," which they define as "deterring an adversary with the threat of unacceptable damage through a retaliatory nuclear strike."[32] The number of scholars who have accepted this characterization suggests that "assured retaliation" may soon become the preferred academic description of Chinese nuclear strategy.[33]

Evidence of transition

Though some scholars find the concepts of minimum deterrence and limited deterrence too restrictive, most still recognize the utility of these terms in creating a common lexicon for Chinese nuclear strategy – at least in the West. This is less the case in China. In fact, a search for "minimum deterrence" and "limited deterrence" in China's most popular academic database, CNKI, reveals only a handful of references over a fifty-year period. The terms are also conspicuously absent from China's Defense White Papers. Without such explicit references in the primary source material, scholars wishing to assess China's strategy must find other means of measurement.

Some scholars have argued, for instance, that it is more accurate to describe Chinese nuclear planning in terms of "nuclear policy" or "nuclear doctrine" than "nuclear strategy," since this is the terminology the Chinese use themselves.[34] The explanation provided for this presumed preference is that the word "strategy" implies nuclear weapons have military utility and are incorporated into the state's defense planning, and that this goes against the fundamental Chinese assumptions regarding nuclear weapons. The Chinese supposedly assume that nuclear weapons are most useful when not used, and thus serve more as political tools than military weapons. Ultimately, scholars making this point do so to provide semantic support for the idea that China subscribes to a strategy of minimum deterrence. The difficulty with this claim, however, is that it lacks empirical support.

A search of the Chinese literature in CNKI dating back to 1988 reveals 4,600 articles which include the word "nuclear strategy," compared to 2,588 articles including "nuclear policy" and 144 articles including "nuclear doctrine." Interestingly, if one breaks these search returns down by year, you see an increase in the use of the word strategy overtime. It is also noteworthy that China's most recent

Table 5.1 Minimum versus limited deterrence

	Minimum Deterrence	*Limited Deterrence*
Objectives	To deter nuclear attack	To deter nuclear and conventional attacks and deny enemy victory
Conditions of use	No First Use	Conditional No First Use/ First Strike Ambiguity
Targets	Heavily populated areas	Enemy military/nuclear forces/ command and control centers
Number of weapons	Enough to inflict significant harm	Enough to have options in nuclear combat
Type of weapons	Survivable, deliverable nuclear systems able to penetrate enemy defenses	Survivable, deliverable, precise, quick-to-launch, short-range nuclear systems able to penetrate enemy defenses + advanced command and control

Defense White Paper, published in 2015, refers to the state's broader "national defense policy" and its nuclear "no first use policy" while also referencing China's "nuclear strategy." If the words in common currency are believed to reveal the underlying assumptions of Chinese nuclear thinking, then evidence indicates that the Chinese *do* consider nuclear weapons to have military utility. This theoretically contravenes minimum deterrence, but to truly make the case that China's nuclear thinking is changing, much more evidence is necessary. The table above outlines the primary differences between minimum and limited deterrence when broken down by objectives, conditions of use, targets, and force requirements. The proceeding discussion evaluates where China falls in these areas.

Objectives

A survey of the public literature produced by the Chinese government over the past half century reveals remarkable consistency in the state's declaratory intent of its nuclear force. Both the state communiques of the late 20th century as well as the Defense White Papers published more recently indicate that the purpose of China's nuclear force is to deter nuclear blackmail and thwart foreign nuclear aggression. One can dismiss such boilerplate language as propaganda, but such dismissals underestimate the utility of mandated consistency in measuring systematic change. Slight deviations in this context have the potential to reveal real shifts in state thinking.

One of the most noticeable alterations of the state position occurred in China's 2008 Defense White Paper, when the paper's authors explained that China's nuclear strategy called for "the flexible use of different means of deterrence." Additional objectives mentioned in the same context included "effectively control[ling] war situations" and "reduc[ing] the risks and costs of war."[35] This language seems to directly contravene the state's public position on the futility and peril of nuclear warfighting, and if it weren't for corroborating military statements, it could likely be seen as an insignificant aberration. The totality of the evidence indicates otherwise. Most critically, several internal Chinese military documents, discussed later, indicate that this statement more accurately reflects the operational aims of China's nuclear force.

China's Defense White Papers provide a top-level overview of its military strategy, situating it within the broader context of the state's peaceful development. It is meant to provide a degree of transparency to the international community as to China's military defense strategy. China's military publications, by contrast, are circulated internally and are meant to provide members of the People's Liberation Army (PLA) with operational details on military strategy. The texts considered most authoritative in this regard are *The Science of Military Strategy*, *The Science of Campaigns*, and *The Science of Second Artillery Campaigns*.

To date, four editions of *The Science of Military Strategy* (SMS) have been published, with an English version published in 2013.[36] These volumes serve as manuals for PLA officers and provide a comprehensive assessment of the strategic direction of the state's military operational planning. As should be expected, the authors of the manual dedicate most of its pages to discussing China's conventional operations, but they nevertheless make clear that China's objective in having a nuclear force centers upon "preventing enemy nations from using or threatening to use nuclear weapons against us."[37]

In addition to *The Science of Military Strategy*, PLA officers also take direction from *The Science of Campaigns* (*SoC*), published by the Chinese National Defense University Press. *The SoC* differs from the *Science of Military Strategy* in its focus on military campaigns rather than doctrine, and thus serves as a valuable supplement in the analysis of the PLA and China's nuclear force in particular. Interestingly, there are places where the *Science of Campaigns* seems to contradict the *Science of Military Strategy*. For example, the *SoC* makes clear that the first mission of China's nuclear force is to deter a foreign nuclear attack, as would be consistent with a strategy of minimum deterrence, but it also delineates the plan of attack if deterrence fails. More specifically, it mentions a counterattack campaign, where the state could engage in "protracted warfare" to contain "the escalation of the nuclear exchange."[38]

This theme continues in the military publication *The Science of Second Artillery Campaigns* (*SSAC*), a classified military training manual intended specifically for China's Second Artillery Force (SAF) officers.[39] Like the *SoC*, the *SSAC* is unequivocal about the necessary expansion of China's nuclear force aims. These aims include deterring foreign nuclear aggression by threatening nuclear retaliation, deterring the escalation of war or "restricting the war to a certain size," and using nuclear weapons as a "backstop" supporting conventional missile strikes.[40]

The last two aims clearly fall outside the parameters of minimum deterrence. They also demonstrate a change in thinking. If nuclear weapons can be introduced to restrict the size and scope of war, then the assumption is that nuclear war can, in fact, be limited. This is also the implication of statements indicating that nuclear weapons "can directly impact the progress and end results of warfare."[41] Moreover, if one intends to use nuclear weapons as a fail-safe behind conventional weapons – with "nuclear missile forces located in the strategic rear" and "conventional missile forces deployed toward the battle front" – then it is assumed that nuclear weapons are not necessarily operationally distinct from conventional weapons but that both weapons can be used together to achieve military aims.[42]

According to the *SSAC*, China's nuclear weapons are not reserved for a single limited retaliatory response, but are meant to *fight* if the occasion calls. This concept is clearly laid out in the manual's introduction. The new era, it says, calls for a shift in the strategic mission of the SAF, "from the single undertaking of guided missile nuclear assault to nuclear and conventional 'dual deterrence and dual operation'."[43] The manual explains further: "Once deterrence has lost its effectiveness, the campaign large formation can quickly transit to actual combat."[44] The presupposition here and in other places where "actual combat" and "actual warfare" are mentioned is that nuclear war-fighting capabilities are a necessary prerequisite to successful deterrence. In this way, deterrence and war fighting are not diametrically opposed but are "interconnected, coexistent, similarly conditioned and closely integrated organic wholes."[45]

This perspective also appears in a Chinese internal military publication *Intimidation Warfare* (*She Zhan*), published by National Defense University and edited by former SAF Lieutenant General Zhao Xijun. *Intimidation Warfare* provides a comprehensive overview of the role of missiles in China's military deterrence at the campaign level, and it clearly echoes many of the points appearing in the *SSAC*, including the aim of nuclear weapons to prevent the escalation of conventional war, the military advantage of using nuclear weapons as a "backing" for conventional forces in battle, and the need to use nuclear weapons for more than psychological deterrence. "In order to contain the escalation of conventional war and maintain the fundamental national interest," writes Zhao, "we must exhibit a strong desire to conduct nuclear retaliation and make the enemy realize that the severe consequences of our nuclear strike exceed the benefits they receive."[46] Later on, Zhao mentions that the goal of the SAF is "prevent[ing] low-intensity nuclear war from further escalating."[47]

Zhao's words and the words in China's other recent military manuals reveal a more nuanced picture than a strategy of minimum deterrence would have one believe. Public statements of some PLA officials provide additional evidence supporting this point. At a seminar on military science in 2000, for instance, PLA major general Dong Qingfu mentioned the increased possibility of nuclear war in the twenty-first century, specifically, the increased possibility of *limited* nuclear war.[48] More recently, PLA rear admiral Yin Zhuo emphasized that China's new SSBN force allows the country to "effectively deter and fight back against those who want to launch nuclear attacks on China."[49] The difference between "strike back" and "fight back" is subtle, but important because it spells out one of the primary differences between minimum and limited deterrence. It also has direct implications for the state's nuclear use policy.

Conditions of use

China is the only nuclear member of the Nuclear Nonproliferation Treaty to officially adhere to a No First Use (NFU) policy. China's Defense White Papers have consistently confirmed this position, with the exception of the 2012 paper, in which the policy did not appear. Though the omission drew international attention, speculation concerning a policy shift was quickly abated as a result of the Chinese government's immediate reassurance of its NFU commitment. The Chinese government also responded by explaining that the omission was a logical consequence of the paper's redesign. In an attempt to shorten the paper, Chinese officials claimed they had changed the paper's focus from "comprehensive," to more "thematic," thereby necessarily truncating the discussion of China's nuclear force – and leaving out China's NFU policy.[50] Closer analysis casts doubt on this explanation.

It is true that China's Defense White Papers had become exceedingly lengthy by 2010, with the number of sections doubling since its first publication. In addition to having a section on the country's national defense, for instance, the 2010 paper also includes separate sections on China's defense expenditures, its military legal system, its defense science and technology, as well as its arms control and disarmament efforts. The 2012 paper omits all of these. It does not, however, leave out the Second Artillery Force altogether. Instead, the 2012 paper includes its discussion of the SAF in section II, where it details the mission of each of its military branches. Moreover, the details provided about the SAF in the 2012 paper were very similar to the ones provided in earlier white papers. For example, the 2012 paper describes the SAF as a "lean and effective" missile force "primarily responsible for deterring other countries from using nuclear weapons against China and carrying out nuclear counterattacks." Unlike other papers, however, it does not go on to specify the *conditions* under which such counterattacks are acceptable. In 2010, this was clarified with one sentence: "China consistently upholds the policy of no first use of nuclear weapons, adheres to a self-defensive nuclear strategy, and will never enter into a nuclear arms race with any other country."

The brevity argument is understandable for a section, but not a sentence. If China's nuclear forces are discussed at all, one would expect inclusion of its *sui generis* nuclear use policy, especially considering the doubt some cast on the credibility of this assurance. It is also highly unlikely that the omission was due to an oversight as some claim, since the Chinese government is well known to heavily vet and scrutinize its official publications. In this case, the paper also likely underwent an interagency review process before reaching the State Council's Information Office for public distribution.[51] It is thus much more likely that the omission was not an oversight, but an intentional signal that echoed at the most official levels what has been observed elsewhere for years – the possibility that China would launch its nuclear weapons before absorbing a first strike or that it would do so after absorbing a *non*-nuclear attack.

Ostensibly, China's No First Use policy requires it to absorb a first strike before engaging in a counterattack. As the *Science of Second Artillery Campaigns* and *The Science of Campaigns* make clear, this would likely require China to respond after sustaining a reduction in its nuclear forces, damage to transport routes, and possibly the destruction of its military command and control centers. It is for this reason that the *Science of Military Strategy* presents another option. The 1999 *SMS* advises, "When it has been judged reliably that the enemy's nuclear weapons are already on their way, but have not yet exploded, respond swiftly by launching a nuclear counterstrike against the enemy."[52] Similar though more detailed guidance appears in the 2013 *SMS*, which states that a counterstrike is appropriate "when it has been clearly determined that the enemy has already launched nuclear missiles against us but said enemy nuclear warheads have yet to arrive at their targets and effectively explode or cause actual damage to us." The manual goes on to explain that these instructions do not directly contravene the country's NFU policy, but serve the greater purpose of ensuring an effective deterrent by mitigating the damage done from an impending attack and improving the chances of a successful counter strike.[53] In reality, though, most scholars and policymakers would agree that these instructions push for action beyond what is proscribed by the state's NFU policy.

So-called "Launch under Attack" or "Launch on Warning" policies are generally seen as inconsistent with the traditional concept of minimum deterrence, but nevertheless, the ex-ante attack is still nuclear. In this way, these policies represent only a slight departure from No First Use, providing a state with greater flexibility as to the *timing* of use, but not releasing the state from the requirement that the precipitating attack be nuclear. Because the objective remains to deter nuclear aggression by threatening a nuclear response, one could theoretically reconcile such policies with the spirit of minimum deterrence. This is not the case when a state expands the conditions of its nuclear use to include instances of non-nuclear aggression as seen elsewhere in the Chinese literature.

One of the most explicit sources on this point is *The Science of Second Artillery Campaigns*, which includes a section entitled "Reducing the Nuclear Deterrence Threshold." In it, the editors list several possible scenarios in which China might consider nuclear use other than responding to a nuclear attack. These include responding to conventional air raids on major Chinese cities or political and economic centers, thwarting the impending victory of an enemy in conventional war, and responding to a conventional attack on China's nuclear or strategic assets. Other sources have also mentioned these situations as possible contingencies of Chinese nuclear use. We will look at them in turn.

The circumstance most often mentioned as potentially compelling a Chinese nuclear response is a large-scale conventional attack. In particular, as conventional missiles become more powerful and their launch-to-strike time shortens, conventional weapons pose an increasing threat to China's critical military, political, and economic assets. To fully protect these areas, China may need to use nuclear weapons to deter conventional aggression. Chu Shulong, the director of the Institute of Strategic Studies at Tsinghua University, is among the scholars who share this perspective. Chu stresses that under normal circumstances China's NFU policy will prevail, but also admits, "If foreign countries launch a full-scale war against China and deploy all types of advanced weapons except nuclear weapons, China may renounce this commitment at a time when the country's fate hangs in the balance."[54] This scenario is equivalent to the conventional victory scenario described in the *SSAC*. The emphasis here is on the potential *scale and scope* of the attack rather than on whether or not the weapon responsible for the attack is nuclear.[55]

But what if the state's survival is not at risk? What if conventional weapons are used in a *limited attack?* Can the threshold for use be lowered still? The executive director of the China Military Science Society suggests it can, saying that nuclear use is an option in situations where "China's security and core interests are under threat."[56] The director does not specify that China must be under nuclear threat, nor does he clarify what constitutes a core interest. On this point, others have weighed in. At least three high-ranking Chinese military officials, for instance, have suggested that China would likely abandon its NFU policy in a conflict with the U.S. over Taiwan. Here, it is less about the type of enemy aggression (nuclear or conventional), and more about the *target* of such aggression and its perceived value.

More recently, the conversation has turned to the possibility of conventional weapons aimed at China's nuclear forces. Shen Dingli, a professor at Fudan University, explains that in this situation too, China could be justified in abandoning its NFU policy.[57] The target is simply too valuable. A similar statement appeared in *Ta Kung Pao* in an interview with military commentator Li Yunsheng. Li mentions that China's nuclear force would strike back only upon a nuclear attack, but he also mentions that certain countries possess precision-guided weapons and that "if China's nuclear facilities are attacked by such weapons, it actually implies the outbreak of a nuclear war, hence the country [China] will consider the use of nuclear weapons under such circumstances."[58] Once again, it is emphasized that the target is too valuable for China to allow for its destruction. Evidence indicates that such arguments are not peripheral, but are taking place at the highest levels in China, with a growing contingent of scholars and experts pushing the Chinese military and government to consider such exceptions.[59] Xu Weidi labels this "conditional no first use," and he argues that its acceptance would effectively mean the end of minimum deterrence.[60] To the extent that a state considers expanded options for nuclear use, then it is accepting the logic of limited deterrence. The same goes for expanding target options.

Targeting

China's No First Use policy explains the conditions under which China would launch a nuclear attack, but it does not specify the necessary nature of the target.[61] This is a separate issue and requires a separate policy, though to date, no state has been inclined to construct one. Instead, the targeting practices of states are largely left to inference, and the guiding assumption is that states subscribing to minimum deterrence limit themselves to targeting civilian areas (counter-value targeting) while states subscribing to limited deterrence also target military assets (counter-force targeting). This pairing makes sense when one considers the discrepant objectives of the two strategies.

States subscribing to minimum deterrence are more likely to threaten to strike high value enemy targets like population centers in response to a nuclear attack, because this approach maximizes the effect of a minimal number of nuclear weapons. In theory, states with even a few nuclear weapons are capable of taking enemy cities hostage and deterring enemy aggression, due to the comparative advantage nuclear weapons have over conventional weapons in their capacity for destruction and the fear they invoke. It is believed that no leader is likely to risk waging a nuclear attack if he conceives there is a possibility that the attack could be met with a nuclear counterattack causing massive suffering and loss of life among his own civilian population. Rational leaders, it is assumed, would not accept this as a cost of war. This assumption anchors minimum deterrence, and there is no contingency for possible exceptions.

Under limited deterrence, by contrast, more weapons make for more options, and a state prepares to both deter nuclear war and to engage in nuclear war if deterrence fails. The latter objective is what necessitates additional targets. After all, the best way to thwart enemy victory or to prevent the escalation of conflict is to destroy or significantly hinder the means by which the enemy can fight back, such as his missiles, troops, and command centers.

China's defense white papers are silent on this point as are generally the Chinese representatives that participate in the Track Two U.S.-China dialogues on nuclear dynamics.[62] The one exception occurred in 2005, when a representative from China admitted that Chinese strategy allowed for targeting enemy military bases and political centers. Curiously, the individual considered his description consistent with counter-value targeting, but, as China's military manuals make clear, counterforce targeting is a more appropriate characterization.[63] The 1999 *Science of Military Strategy*, for instance, describes China's nuclear mission as "crippl[ing] enemy war potential and strategic strike forces, retard[ing] its strategic operations, sabotag[ing] its strategic intentions, and caus[ing] psychological shock."[64] The most recent edition also refers to strikes that "have a major impact on the broader strategic picture."[65] The 2013 *Science of Campaigns* employs similar language, tasking the country's nuclear force with impeding the enemy's war potential, strategic intentions, and determination.[66]

Neither the *SMS* nor the *SoC* mention the specific targets necessary to achieve their objectives; they leave this to *The Science of Second Artillery Campaigns*. According to the *SSAC*, retaliation requires targeting "an enemy's command centers, communication hubs, military bases, political centers, economic centers, important industrial bases, and other strategic campaign targets."[67] The second chapter of *Intimidation Warfare* includes similar language. "Strategic missile forces can collaborate with other nuclear troops in operation," it states,

> and can also independently implement strategic nuclear assault, mainly striking strategic targets such as the enemy strategic nuclear weapons base, strategic rear-area base, state and military command headquarters, and political and economic centers, heavy industry facilities, transportation and communication nexus and heavy military industry groups.[68]

The text later specifies that China's submarine launched ballistic missiles, in particular, are weapons suitable to accomplish this mission, since they are able to "independently implement nuclear assaults, striking the enemy's important political, economic, and military strategic targets."[69]

Nowhere in *Intimidation Warfare* or the *SSAC* do the authors mention the strategic advantage of targeting population centers. Instead, deterrence is best accomplished by threatening the destruction of physical assets. The *SSAC* makes this clear when it equates the success of an SAF campaign with the amount of physical damage inflicted on a specified target. If a target is not sufficiently destroyed, then the manual instructs commanders to immediately commence a second wave of attacks.[70] The goal is not solely to inflict psychological damage but to actually render enemy combat capabilities impotent.

Of course, the realization of such strategic goals and targets requires the necessary capabilities. Twenty years ago, when Johnston first introduced the idea of limited deterrence, he claimed China did not have the weapons to make this a reality. Today, China is in a much better position to do so.

Force structure

The most recent assessment of China's nuclear force estimates that it has 280 operational nuclear weapons.[71] The question of whether this constitutes enough weapons to accomplish the strategic objectives of limited deterrence depends largely upon the *types* of weapons this includes. More specifically, a strategy of limited deterrence requires certain capabilities, including mobile missiles which can launch quickly and strike small targets and missiles which can strike targets at short range.

Over the past twenty years, China has upgraded both its land- and sea-based missiles. One of its earliest advancements was to enable its land-based missiles to run on solid fuel as opposed to liquid fuel. This presents a two-fold advantage. First, it shortens launch time. The highly caustic nature of liquid fuel necessitates that rockets are fueled shortly before launch. Using solid fuel eliminates this step, because the fuel is more stable and built into the weapon itself. Second, the size of the weapon can be reduced. To accommodate liquid fuel, missiles must be larger in size. Solid fuel, however, is more compact, allowing for smaller missiles which can then allow for greater mobilization.

China's switch from liquid-fueled to solid-fueled missiles raised little suspicion from foreign powers, because it was seen as a necessary step in ensuring the survival of China's missiles. Moreover, since force survivability is the *sin qua non* of minimum deterrence, these developments appeared generally consistent with China's strategy of minimum deterrence. After all, minimum deterrence requires a state to present a credible threat of counteraction and this threat is significantly undermined if an aggressor can easily locate and target a state's missiles. For minimum deterrence, uncertainty is paramount. Potential attackers must doubt that a first strike will eliminate all missiles. They must fear the possibility that some weapons will survive a first strike and be able to retaliate. By putting missiles on mobile launchers as opposed to in stationary silos, a state makes it more difficult for an enemy to accurately assess the location of its missiles at any given time – and thus also makes it more difficult for an enemy to know with certainty that a first strike will not be met with a counterstrike.

The other advantage of switching fuel types is that it enables missiles to be launched more quickly, but this is really only an advantage in certain scenarios. Clearly, if your strategy requires that you wait to absorb a first strike before striking back, as does minimum deterrence, then the few minutes spent fueling missiles is not your concern. If, however, a state considers the option of launching its weapons upon warning of an attack or launching them upon confirmation of an incoming attack, as is done under limited deterrence, then delays become costlier. The fact that some Chinese reports boast about this capability leaves room for speculation.[72] Military expert Cao Weidong, for instance, stated that "China's strategic nuclear submarines can launch a counterattack when they themselves are under attack, thus serving as deterrent to the enemy use of nuclear weapons."[73] The problem, of course, is that this capability doesn't align with China's commitment to No First Use. Another report is more direct: "the mobility, range, precision, and warhead yield [of the DF-31] combine to give it formidable first-strike nuclear capability."[74] Why highlight the quickness of launch if your nuclear policy makes it irrelevant? And why mention the enhanced ability to strike first if the state's nuclear policy specifically forbids it?

Similar questions can be asked of the increased precision of China's nuclear missiles, but, in this case, one could ask not only why China is communicating this force advancement, but why China is improving precision of its missiles at all. The answer goes to a likely change in China's targeting practices, as explained above. If cities are your intended target, as minimum deterrence suggests, then increased missile accuracy is relatively moot. Accuracy becomes very important, though, when a state wants to strike military targets, which are smaller and more mobile.[75] China's newer weapons make this a greater possibility.

China's first land-based nuclear weapon, the Dongfeng-3A had a Circular Error Probable (CEP), a measure of a weapon system's precision, of one to four kilometers. This means that at least 50 percent of the time, the warhead could strike a target within a radius of one to four kilometers. By contrast, China's most recent long-range missile, the Dongfeng-41 has an estimated CEP of 100–200 meters.[76] China's intermediate-range weapons and sea-based nuclear weapons have also become more accurate, with the Dongfeng-26 having an estimated CEP of 150–450 meters (as opposed to the earlier Dongfeng-4 with a CEP of 1.5 kilometers) and the sea-based Julang-2 being more than twice as accurate as its predecessor, with an estimated CEP of 150–300 meters. Gone are the days, says one reporter, when a Chinese missile is characterized by "a low accuracy rate, a big size, a heavy weight, difficult technological support, long-term war preparations, a high campaign tactical cost, and poor results."[77] Though these weapons would be suitable to meet the needs of minimum deterrence, China has opted to modernize its force and add to its capabilities. Its missiles now are more precise, smaller, lighter, more mobile, and better supported. Most recently, China has also made some of its missiles more efficient.

In 2010, it was confirmed that China had retrofitted the Dongfeng-5 to accommodate multiple warheads.[78] In 2012, China tested its most advanced intercontinental ballistic missile (ICBM), the Dongfeng-41, which has the capacity to carry up to ten warheads. This missile, tested again in 2017 and 2018, is expected to be commissioned in 2019.[79] In military terms, this means that China has at least two systems which can be considered multiple independently targetable reentry vehicles (MIRVs).[80] This designation is reserved for missiles that are designed to carry multiple miniaturized warheads, each equipped with its own electronic guidance system and radar. Once a "MIRVed" missile loses its spent boosters and nose cone, it releases the warheads one at a time to strike different targets. MIRVing the Dongfeng-5 and Dongfeng-41 provides China with two advantages: it increases the number of warheads per missile and it increases the number of targets available per missile. If one emphasizes the former, then China's MIRVed missiles can be seen as consistent with a strategy of minimum deterrence.

Survivability is only a necessary condition for states subscribing to minimum deterrence. It lends credibility to their threat of retaliation, but it is not alone enough to deter a first strike. In addition to demonstrating that some of its nuclear missiles are likely to remain after a first strike, a state under minimum deterrence must also show that its remaining weapons are capable of delivering a second strike. In some cases, this means that China must have a chance at getting past a country's missile defenses.

The U.S. is the only country to have a comprehensive missile defense system intended to protect the entire continental United States. From the perspective of China, the U.S. system thus represents a threat to China's second-strike capability and destabilizes the U.S.-Sino deterrence relationship. The U.S. missile defense system is currently designed to intercept incoming missiles during the mid-course phase of their flight trajectory (the 19–20 minutes when the missile has reached its highest altitude). Since this is the point when warheads detach from their carrier missile, successful defense requires the destruction of the warheads as opposed to the missile. One of the technological limitations of the U.S. system is that it takes at least one anti-ballistic missile to destroy one warhead.[81] This means that if an incoming missile carries multiple warheads, then multiple anti-ballistic missiles are necessary to destroy them. If there are a limited number of interceptors (as is the case with the U.S.), then they are more easily overwhelmed, and it is more likely that a warhead will get through to strike its intended target. This likelihood is compounded when one considers that each warhead has its own navigation system and target. Consequently, many scholars and policymakers believe China has MIRVed its missiles in response to the increased number of U.S. anti-ballistic missiles.[82] There are at least two reasons, however, why MIRVs do not automatically align with a strategy of minimum deterrence.

The first reason is that by MIRVing one's missiles, a state sends a signal to other states regarding the increased ease at which it could surge its number of deliverable warheads. Since minimum deterrence deemphasizes quantitative gains in arms, this ability seems moot, at best. A better approach to counter missile defense, then, would likely be to gradually develop more missiles, or, better yet, to develop warhead decoys. This approach, which China appears to have taken in the past, causes less alarm. The second reason MIRVs make less sense for states subscribing to minimum deterrence is because of the inherent risk they pose to force survivability. Just as MIRVs increase the chances of defense penetration, they can also *decrease* chances of weapon survival by creating an incentive for an enemy state to strike first. By striking grounded enemy missiles loaded with multiple warheads, potential aggressors can more efficiently wipe out target forces, using fewer missiles to eliminate more warheads.

The ability of a state to quickly increase the number of its deliverable warheads seems well suited for a strategy that emphasizes greater force levels, like limited deterrence. Moreover, under limited deterrence the destabilizing feature of MIRVs is offset by the strategic advantage they can provide in a war scenario. MIRVs enable states to strike multiple targets with one missile. This can be ideal in a war situation where one is striking several smaller targets, but it is less critical if one's strategy calls for a single retaliatory strike on a large stationary target. In the case of the latter, the ability to target multiple targets at once can be considered "overkill," as one Chinese report admits.[83]

A similar critique can be made concerning Chinese nuclear testing. China ceased live nuclear testing in 1996 after signing the Comprehensive Nuclear Test Ban Treaty (CTBT), but its *simulated* tests continued.[84] In fact, though China has only gradually added new systems to its nuclear force, it has simulated more nuclear tests over the past five years than any other country in the world.[85] One explanation proffered for China's increased level of testing is simply its desire to "catch up" to the U.S. technologically, which had the opportunity to conduct over twenty times more live nuclear tests than China before signing the CTBT. Another reason includes the development of a new generation of smaller and more precise nuclear weapons. In either case, most concur that such aggressive testing signals that China will continue to upgrade its force in the coming years to increase both the size and sophistication of its nuclear arsenal.[86]

Conclusion

The size of China's nuclear force relative to the two nuclear superpowers can lead to an understated appreciation of the country's nuclear status. The same can be said for looking at China's force buildup. While China is the only NPT (Treaty on the Non-Proliferation of Nuclear Weapons)-nuclear weapon state to continue to increase its nuclear force levels, its growth has been gradual and is often characterized as one of restraint. China's force modernization, however, is less easily dismissed. Over the past decade, China has deployed new nuclear delivery systems for every part of its nuclear triad, including new mobile intercontinental ballistic missiles, new submarine-launched ballistic missiles, and a new bomber that can carry nuclear cruise-missiles. The nuclear weapons it now has in development are even more impressive. These include the MIRVed Dongfeng-41 ICBM and a potentially MIRVed SLBM, the Julang-3.[87] China is also distinguishing itself by testing anti-ballistic missiles, anti-satellite weapons, and hypersonic weapons.[88]

As China continues to increase its nuclear capabilities, it becomes increasingly difficult to justify these as necessary to ensure China's second-strike capability. An alternate explanation, supported by China's expanded nuclear objectives, conditions of use, and targets, is that these weapons are meant to increase China's combat capability. More specifically, it appears China wants to have the option to use its nuclear missiles prior to a nuclear strike, possibly in tandem with China's conventional missiles to deter further enemy aggression or interference. This suggests that China is prepared to transition to a strategy of limited nuclear deterrence. In fact, some in China have openly advocated for such strategic change. After a draft of the U.S. *Nuclear Posture Review* was leaked, for instance, the state-run *Global Times* published an editorial suggesting that China should seriously reconsider its " 'minimal reprisal' strategy" in light of new U.S. nuclear developments.[89]

The criticism that quickly followed these comments in other Chinese news outlets highlights the difficulty in accurately assessing the strategic situation.[90] It is important to emphasize that China is much less transparent in regards to its nuclear strategy than the other states explored in this volume, leaving scholars and policymakers to piece together the evidence and come to their own conclusions. This chapter should thus not be construed as to provide a definitive answer. Scholars using the same or similar sources as the ones chosen here can and have reached different conclusions. The scholarly consensus, to date, has favored minimum deterrence, though a growing contingent of scholars continue to speculate that China has greater ambitions for its nuclear force. The present U.S. presidential administration employs the latter perspective.

The 2018 U.S. *Nuclear Posture Review* confirms that the U.S. sees China as taking steps away from minimum deterrence, and it warns China against going any further. In particular, the report states that the U.S. is prepared, "to prevent Beijing from mistakenly concluding that it could secure an advantage through the limited use of its theater nuclear capabilities, or that any use of nuclear weapons, however limited, is unacceptable."[91] The report also warns that "the costs incurred as a result of Chinese nuclear employment, at any level of escalation, would vastly outweigh any benefit."[92] These warnings suggest that the U.S. finds it likely that China's nuclear strategic doctrine may indeed shift away from a strategy of minimal deterrence and may consider the use of nuclear weapons in situations other than in response to a nuclear attack and in other ways than striking enemy cities. It's for this reason the U.S. labeled China a "potential adversary" in the report. China has a similar view.

Though China must manage relations with multiple nuclear neighbors, the consensus among academicians and policymakers is that its nuclear weapons development and strategy is primarily shaped by the United States. This is due to several factors. First is China's nuclear preeminence in the Indo-Pacific region. This mitigates China's concern from longstanding regional rivals like India. Whereas China has both submarine-launched ballistic missiles and intermediate-range ballistic missiles with the necessary range to strike targets well within India, the threat India poses to China is less severe. In fact, the ability of India to successfully strike China at all faces serious constraints. Its nuclear bombers would need to evade Chinese air defenses, and its short-range submarine-launched ballistic missiles would have to survive China's anti-submarine capabilities.[93] India's land-based missiles would likely fare better, but at present, their range and launch position limit the targets available.[94] This is perhaps why one can say that "China is not worried about India at all from a nuclear standpoint" – as one Chinese expert did at a U.S.-China Strategic Dialogue.[95] Another participant at a more recent conference opined, "China knows for certain that nuclear deterrence works well between China and India."[96]

In dyadic regional relationships where China does not have the nuclear advantage, it has longstanding cooperation. Presently, Russia has eight strategic nuclear weapons for every one of China's nuclear weapons and has a solid diversification of nuclear platforms (land, air, and sea). It also leads the world in tactical nuclear weapons and maintains a limited ballistic missile defense system. Looking solely at capabilities and proximity, Russia appears to present the greatest threat to Chinese security. This threat is heavily mitigated, however, by the two countries' repeated positive interactions over the past half century, their shared commitment to noninterference, and their parallel vision of a multipolar world order.[97] Even evidence of Russia's violation of the Intermediate Range Nuclear Force (INF) Treaty seemed not to cause extreme consternation in China.[98] Instead, it was the U.S. withdrawal from the INF that caused alarm. In particular, the Chinese saw the move as one which allows the U.S. to further contain China by developing and deploying medium and intermediate-range ballistic missiles in U.S.-ally territory in the Indo-Pacific region.[99]

If the U.S. intent is truly to contain China, as many in China believe, then the United States can recruit its regional Allies to help it achieve this objective. This is the luxury of a superpower, and it is a large reason why China perceives the United States as its primary security threat. China wants to change this, and to "reorder the region to its advantage."[100] In particular, China wants to enforce its territorial claims in the South China Seas vis-à-vis Japan and its claims in the East China Seas vis-à-vis the Philippines, Malaysia, Vietnam, and Indonesia and deter the U.S. from interfering. It also seeks to use its military might, both conventional and nuclear, to dissuade the U.S. from intervening in any situation having to do with Taiwan. China's new nuclear weapons are instrumental in its pursuit of these objectives as is a nuclear strategy which signals increased flexibility of use. Even if there are those in the U.S. who maintain that the chances of a U.S.-China nuclear confrontation are "somewhere between nil and zero," it is clear from the evidence that Chinese leaders do not have the same level of confidence and are shifting their nuclear strategy accordingly.[101]

Notes

1 Eric Heginbotham, Michael S. Chase, Jacob Heim, Bonny Lin, Mark R. Cozad, Lyle J. Morris, Christopher P. Twomey, Forrest E. Morgan, Michael Nixon, Cristina L.

Garafola, and Samuel K. Berkowitz, *China's Evolving Nuclear Deterrent: Major Drivers and Issues for the United States* (Santa Monica, CA: RAND Corporation, 2017), 2.

2 Brad Roberts, Robert A. Manning, and Ronald N. Montaperto, "China: The Forgotten Nuclear Power," *Foreign Affairs* 79, no. 4 (July–August 2000), 53–63.

3 Alastair Iain Johnston, "China's New 'Old Thinking': The Concept of Limited Deterrence," *International Security* 20, no. 3 (Winter 1995–1996), 5–42.

4 Chu Shulong and Rong Yu label this "dynamic minimum deterrence:" Chu Shulong and Rong Yu, "China: Dynamic Minimum Deterrence," in Muthiah Alagappa (ed.), *The Long Shadow: Nuclear Weapons and Security in 21st Century Asia* (Stanford, CA: Stanford University Press, 2008), 161–187.

5 Seymour Topping, "China Tests Atomic Bomb," New York Times, October 16, 1964.

6 Li Bin and Tong Zhao, *Chinese Nuclear Thinking* (Washington, DC: Carnegie Endowment for International Peace, 2016).

7 John Wilson Lewis and Xue Litai, *China Builds the Bomb* (Stanford, CA: Stanford University Press, 1988).

8 Jeffrey Lewis, *The Minimum Means of Reprisal: China's Search for Security in the Nuclear Age* (Cambridge, MA: American Academy of Arts and Sciences, 2007).

9 Renny Barbiarz, "The People's Nuclear Weapon: Strategic Culture and the Development of China's Nuclear Weapons Program," *Comparative Strategy* 34, no. 5 (2015), 422–446.

10 China's first Defense White Paper, published in 1998, included a No First Use Pledge as well as a commitment not to enter into a nuclear arms race. In 2000, China added a promise not to engage in nuclear war with any nonnuclear country. The 2002, 2004, 2006, 2008, and 2014 Defense White Papers contain similar wording. See: The Information Office of China's State Council, "China's National Defense in 2008," July 1, 1998, Sec. II; Information Office, "China's National Defense in 2002," December 9, 2002, Sec. II; Information Office, "China's National Defense in 2004," December 27, 1998, Chapter X; Information Office, "China's National Defense in 2006," December 2006, Sec. II; Information Office, "China's National Defense in 2008," January 20, 2009, Sec. II & VII; Information Office, "China's National Defense in 2010," March 31, 2011, Sec. II & VII; Information Office, "China's Military Strategy in 2015," May 26, 2015, Sec. IV.

11 Yao Yunzhu, "Chinese Nuclear Policy and the Future of Minimum Deterrence," *Strategic Insights* 6, no. 9 (September 2005).

12 See: Shulong and Yu, "China: Dynamic Minimum Deterrence"; Sun Xiangli, "The Development of Nuclear Weapons in China," in Li Bin and Tong Zhao (eds.), *Understanding Chinese Nuclear Thinking* (Washington, DC: Carnegie Endowment for International Peace, 2016), 79–102.

13 Hui Zhang, "China's Nuclear Weapons Modernization: Intentions, Drivers, and Trends," Belfer Center, Harvard University, July 2012; Gregory Kulacki, "China's Nuclear Arsenal: Status and Evolution," *Union of Concerned Scientists*, October 2011.

14 Johnston, "China's New 'Old Thinking'," 5–42.

15 Ibid., 9.

16 Ibid., 5–6.

17 Ibid., 31.

18 Ibid., 41.

19 Baohui Zhang, *China's Assertive Nuclear Posture: State Security in an Anarchic International Order* (New York: Routledge, 2015); Susan Turner Haynes, *Chinese Nuclear Proliferation: How Global Politics Is Transforming China's Weapons Buildup and Modernization* (USA: Potomac Books and University of Nebraska Press, 2016); James Samuel Johnson, "Chinese Evolving Approaches to Nuclear 'War-Fighting': An Emerging Intense U.S.-China Security Dilemma and Threats to Crisis Stability in the Asia Pacific," *Asian Security*, March 2018; Anthony H. Cordesman, "China and the

New Strategic Nuclear Arms Race: The Forces Driving the Creation of New Chinese Nuclear Delivery Systems, Nuclear Weapons, and Strategy," Center for Strategic and International Studies, November 15, 2018; Mark B. Schneider, "Nuclear Weapons in Chinese Military Strategy," No. 441, National Institute of Public Policy, National Institute Press Information Series, May 3, 2019.

20 Zhang, *China's Assertive Nuclear Posture*; Johnson, "Chinese Evolving Approaches to Nuclear 'War-Fighting'"; Dean Cheng, "Evolving Chinese Thinking About Deterrence: The Nuclear Dimension," The Heritage Foundation, August 16, 2017, www.heritage.org/asia/report/evolving-chinese-thinking-about-deterrence-the-nuclear-dimension.

21 Michael Mazza and Dan Blumenthal, *China's Strategic Forces in the 21st Century: The PLA's Changing Nuclear Doctrine and Force Posture* (Arlington, VA: Nonproliferation Policy Education Center, April 6, 2011), http://npolicy.org/article.php?aid=813&rt=&key=michael %20mazza&sec=article&author=.

22 Office of the Secretary of Defense, Annual Report to Congress: Military and Security Developments Involving the People's Republic of China 2017, Department of Defense, May 15, 2018, 59.

23 Ibid., 6.

24 A secondary benefit is the threat the DF-26 poses to India.

25 This is made clear in the DOD report, when it mentions that China's strategic interest with its intermediate range ballistic missiles (IRBMs) is to secure counter U.S. power projection by threatening U.S. air bases, logistics and port facilities, communications, and other ground-based infrastructure in the region (p. 60).

26 James M. Acton, "Is China Changing Its Position on Nuclear Weapons?" New York Times, April 18, 2013.

27 Mazza and Blumenthal, *China's Strategic Forces in the 21st Century*.

28 Cheng, "Evolving Chinese Thinking About Deterrence." The Heritage Foundation, 2017, p. 4.

29 Michael Swain, "China and Weapons of Mass Destruction: Implications for the United States," Conference Report, National Intelligence Council, November 5, 1999.

30 James C. Mulvenon, Murray Scot Tanner, Michael S. Chase, David Frelinger, David C. Gompert, Martin C. Libicki, and Kevin L. Pollpeter, *Chinese Responses to U.S. Military Transformation and Implications for the Department of Defense* (Santa Monica, CA: RAND Corporation, 2006), 97.

31 Bates Gill, James Mulvenon, and Mark Stokes, "The Chinese Second Artillery Corps: Transition to Credible Deterrence," in James C. Mulvenon and Andrew N. D. Yang (eds.), *The PLA as Organization: Reference Volume v1.0* (Santa Monica, CA: RAND, 2002), 549.

32 Taylor Fravel and Evan S. Mederios, "China's Search for Assured Retaliation: The Evolution of Chinese Nuclear Strategy and Force Structure," *International Security* 35, no. 2 (2010), 48–87.

33 See: Frank G. Klotz and Oliver Bloom, "China's Nuclear Weapons and the Prospects for Multilateral Arms Control," *Strategic Studies Quarterly* 7, no. 4 (2013), 3–10; Vipin Narang, *Nuclear Strategy in the Modern Era: Regional Powers and International Conflict* (Princeton, NJ: Princeton University Press, 2014); Fiona S. Cunningham et al., *Assuring Assured Retaliation: China's Nuclear Posture and U.S.-China Strategic Stability* (Cambridge, MA: MIT Press, 2016); Kenneth W. Allen and Jana Allen, "Building a Strong Informatized Strategic Missile Force: An Overview of the Second Artillery Force with a Focus on Training in 2014," *The China Brief*, Jamestown Foundation, April 2016.

34 Yunzhu, "Chinese Nuclear Policy and the Future of Minimum Deterrence"; Monika Chansoria, *Nuclear China: A Veiled Secret* (Daryaganj and New Delhi: K W Publishers Pvt Ltd, 2014).

35 Information Office of the State Council of the People's Republic of China, *China's National Defense in 2008*, Beijing, January 2009, Sec. II, www.china.org.cn/government/whitepaper/node_7114675.htm.

36 In specific regards to the latest version of the *SMS*, published in English, it is wise to heed the advice of Andrew Erickson: "Since this book has deliberately been made accessible to an overseas audience, it is important to reflect on what message its English-language publication may be intended to convey." Andrew S. Erickson, "The Science of Military Strategy," *Naval War College Review* 60, no. 3 (2007), Article 11, http://digital-commons.usnwc.edu/nwc-review/vol60/iss3/11.
37 Gregory Kulacki, *The Chinese Military Updates China's Nuclear Strategy* (Cambridge, MA: Union of Concerned Scientists, March 2015), 1.
38 Houqing Wang and Xingye Zhang, *Zhanyi Xue* [*The Science of Campaigns*] (Beijing: National Defense University Press, 2000), 372.
39 The Second Artillery Force changed names to the PLA Rocket Force in 2016, when it became a full-service equivalent to the PLA Army, Navy, and Air Force. SAF is used in this text because this is the name referenced in the manuals mentioned.
40 Jixun Yu, *Dier Paobing Zhanyixue* [*The Science of Second Artillery Campaigns*] (Beijing: PLA Press, 2004), 274.
41 Ibid., 55.
42 Ibid., 118.
43 Ibid., 12.
44 Ibid., 274.
45 Ibid., 126. This is not to imply that deterrence is achieved through war fighting but that it is achieved through the demonstration of a state's war-fighting capabilities. The *SSAC* mentions that this can be done through signaling, intimidation, propaganda, test launches, military exercises, and counter-nuclear deterrence.
46 Xijun Zhao, *She Zhan: Daodan Weishe Zonghengtan* [*Intimidation Warfare: A Comprehensive Discussion on Missile Deterrence*] (Beijing: National Defense University, 2005), 86, 83, 31.
47 Ibid., 40.
48 Xinhua, "*PLA Daily* Carries Article on Guarding Against Nuclear War," April 18, 2000.
49 Xinhua, "China Showcases Its Nuclear Submarine Force," October 29, 2013.
50 Hui Zhang, "China's No-First-Use Policy Promotes Nuclear Disarmament," *The Diplomat*, May 22, 2013.
51 Michael Kiselycznyk and Phillip C. Saunders, *Assessing Chinese Military Transparency*, Institute for National Strategic Studies China Strategic Perspectives, No. 1 (Washington, DC: National Defense University Press, June 2010), 4.
52 Wang Wenrong, ed., *Zhanlue xue* [*The Science of Strategy*] (Beijing: Guofang Daxue Chubanshe, 1999).
53 Michael S. Chase, "Nuclear Policy Issues in the 2013 Edition of the Science of Military Strategy: Part I on Nuclear Policy, Strategy, and Force Modernization," *China Brief* 15, no. 11 (2015), 4–8.
54 Chu Shulong, "China Should Maximize the United States and Japan to Deal with the Threat of Nuclear Deterrence," *Renmin Ribao*, June 4, 2014.
55 This statement makes more sense when one considers the modern advancements in conventional missiles exemplified by U.S. Conventional Prompt Global Strike, where long-range powerful missiles are intended to reach their target in mere minutes. The same rationale could be applied to justify the use of nuclear weapons in response to a devastating cyber-attack.
56 Meng Ching-shu, "PRC Expert on China's Nuclear Counterattack Principle in White Paper," *Wen Wei Po*, April 17, 2013.
57 Shen Dingli, "Nuclear Deterrence in the 21st Century," *China Security* 1, no. 1 (Summer 2005), 10–14.
58 Shang Yi, "Zhu Chenghu: Foreign News Agency 'Cites Out of Context'," *Ta Kung Pao*, July 16, 2005.

59 Larry M. Wortzel, *China's Nuclear Forces: Operations, Training, Doctrine, Command, Control, and Campaign Planning* (Carlisle Barracks, PA: U.S. Army War College, Strategic Studies Institute, 2007), 27.
60 Xu Weidi, "China's Security Environment and the Role of Nuclear Weapons," in Li Bin and Tong Zhou (eds.), *Chinese Nuclear Thinking* (Washington, DC: Carnegie Endowment for International Peace, 2016).
61 The only targeting aspect of NFU is that China promises not to target non-nuclear countries.
62 Track 2 Dialogues represent an unofficial forum for U.S. and Chinese experts and sometimes also government officials to discuss U.S.-Sino nuclear dynamics. The dialogues take place annually and are organized by the Pacific Forum, Center for Strategic and International Studies, Center for Contemporary Conflict, Naval Postgraduate School, and Defense Threat Reduction Agency.
63 Michael Glosny and Christopher Twomey, *U.S.-China Strategic Dialogue, Phase V* (Washington, DC: U.S. Naval Postgraduate School and Pacific Forum of the Center for Strategic and International Studies, 2010), 84.
64 Wang and Zhang, *Zhanyi Xue*, 355.
65 Chase, "Nuclear Policy Issues," 7.
66 Wang and Zhang, *Zhanyi Xue*, 369.
67 Yu, *Dier Paobing Zhanyixue*, 241.
68 Zhao, *She Zhan*, 17.
69 Ibid., 18.
70 Yu, *Dier Paobing Zhanyixue*, 307.
71 Hans M. Kristensen and Robert S. Norris, "Chinese Nuclear Forces, 2018," *Bulletin of the Atomic Scientists* 74, no. 4 (2018), 289–295.
72 He Tianjin, "A Second Artillery Brigade Taps Latent Combat Power," Renmin Ribao, August 20, 2003.
73 Sun Xiangli, "China's Nuclear Strategy: Nature and Characteristics," *World Economics and Politics*, no. 9 (2006), 26.
74 "Donfeng-41: China's New Intercontinental Ballistic Missile," *CNTV*, August 29, 2012, http://english.cntv.cn/program/china24/20120829/101673.shtml.
75 Another advantage is the way in which China's increasingly precise medium-range ballistic missiles and intermediate range ballistic missiles can be used as derivatives for its satellite and ballistic missile interceptors.
76 This is comparable to the U.S. Minuteman III ICBM and Russian Yars RS-24, which both have a CEP of 150–200 meters.
77 "Special Dispatch: 'Aces' in 'Dongfeng' Family – Miniaturization, Solidification, and Mobility," *Ta Kung Pao*, October 2, 1999.
78 Though the Pentagon was the first to confirm China's MIRVed missiles, it is widely believed that China has had this ability since the 1990s. This belief derives from multiple demonstrations in the preceding decade where China launched several independently orbiting satellites from one rocket.
79 Andrew Tate, "China Moves Closer to Commissioning DF-41 ICBM After Latest Test Launch," *Jane's Defence Weekly*, June 13, 2018.
80 The other system that could possibly be MIRVed is the Dongfeng-31. Suspicions of a new, MIRVed DF-31 emerged as a result of the PLA displaying an improved launcher, the DF 31AG, in its 90th Anniversary Parade in 2017. These suspicions have yet to be substantiated.
81 In actuality, it is more likely that the U.S. will launch four ABMs to try to destroy one incoming missile, according to the Director of the Missile Defense Agency in his testimony before Congress. See: Status of Implementing the Phased Adaptive Approach to Missile Defense in Europe, Hearing Before the Subcommittee on Strategic Forces of the Committee on Armed Services, House of Representatives, 111th Congress, 33.

Statements by MDA Director Lt. Gen. Patrick O'Reilly and House Representative Trent Franks of Arizona, 2010, www.gpo.gov/fdsys/pkg/CHRG-111hhrg65294/pdf/CHRG-111hhrg65294.pdf.

82 This explanation assumes that the U.S. can and will improve the accuracy of its ABMs, as well as add additional ABMs to its force. At present, the U.S. has 33 ground-based midcourse defense interceptors, with the Missile Defense Agency testifying that four interceptors are recommended for the destruction of one warhead. China has approximately 35 missiles that can deliver a nuclear warhead to the continental United States.

83 Yun Zhen, "Three Nuclear Submarines Launched Guided Missiles to Accurately Hit Targets," Renmin Ribao, August 31, 2001.

84 China's adherence to the CTBT is de facto, since neither the U.S. nor China have ratified the treaty.

85 Stephen Chen, "China Steps Up Pace in New Nuclear Arms Race with U.S. and Russia as Experts Warn of Rising Risk of Conflict," *South China Morning Post*, May 28, 2018.

86 Ibid.

87 Bill Gertz, "China Flight Tests Submarine Launched Ballistic Missiles," *The Washington Free Beacon*, December 18, 2018, https://freebeacon.com/national-security/china-flight-tests-new-submarine-launched-missile/.

88 With its first ASAT test in 2007 and its first ABM test in 2010, China was the third country, following the U.S. and Russia, to test such technology. Its ballistic and anti-satellite interception capabilities now include the DN-3, DN-2, the HQ-19, and the SC-19 designed for anti-satellite missions. China's hypersonic advances were showcased with seven experimental HGV tests of the DF-ZF between 2014 and 2016. In 2017, China tested a MRBM equipped with an HGV (the DF-17) meant for operational use.

89 "The United States Strengthens Its Nuclear Superiority, and China Cannot Afford to Ignore it," *Global Times*, January 9, 2018, http://opinion.huanqiu.com/editorial/2018-01/11506799.html.

90 See Raymond Wang, "Making Sense of Chinese Reactions to the U.S. 2018 Nuclear Posture Review," *The Diplomat*, February 27, 2018.

91 U.S. Department of Defense, *Nuclear Posture Review Report* (Washington, DC: U.S. Dept. of Defense, 2018, 30), https://media.defense.gov/2018/Feb/02/2001872886/-1/-1/1/2018-NUCLEAR-POSTURE-REVIEW-FINAL-REPORT.PDF.

92 Ibid., 32.

93 India's only operational SLBM has a range of only 400 km, requiring it to travel outside Indian waters to strike China.

94 India's IRBMs, the Agni-3 and Agni-4 have ranges of 3,200 km and 3,500 km, respectively, making them capable of striking many of China's cities, but not those on its Eastern Coast like Beijing and Shanghai. Their reach and potential targets are limited further when positioned away from the Sino-Indian border, which is necessitated by the threat posed by China's conventional ballistic missiles. See: Masahiro Kurita, "China-India Relationship and Nuclear Deterrence," *NIDS Journal of Defense and Security* 19 (December 2018), 37–61.

95 Quoted in Ralph A. Cossa, Brad Glosserman, and Matt Pottinger, "Progress Despite Disagreements: The Sixth China – U.S. Strategic Dialogue on Strategic Nuclear Dynamics," *Issues & Insights* 12, no. 5 (November 2011), 9, http://csis.org/publication/issues-insights -vol-12-no5.

96 Michael Glosny, *U.S.-China Strategic Dialogue Phase VIII Report, 2014–008* (Monterey, CA: Naval Postgraduate School and the Pacific Forum Center for Strategic and International Studies, 2014), 6, https://calhoun.nps.edu/handle/10945/44733.

97 For an extended discussion on these points see: Susan Turner Haynes, "China's Nuclear Threat Perceptions," *Strategic Studies Quarterly* 10, no. 2 (2016), 25–62.

98 Though China seems generally unconcerned about the Russian nuclear threat, the same cannot necessarily be said for Russia vis-à-vis China. China's new IRBMs and Ground Launched Cruise Missiles (GLCMs) are among the reasons cited by Russia as justifying its abrogation of the INF Treaty.

99 Tong Zhao, "Why China Is Worried About the End of the INF Treaty," *Carnegie-Tsinghua Center for Global Policy*, November 7, 2018, https://carnegietsinghua.org/2018/11/07/why-china-is-worried-about-end-of-inf-treaty-pub-77669.

100 Office of the Secretary of Defense, *Missile Defense Review, 2019: P.V.*, https://media.defense.gov/2019/Jan/17/2002080666/-1/-1/1/2019-MISSILE-DEFENSE-REVIEW.PDF.

101 Dennis C. Blair and Caitlin Talmadge, "Would China Go Nuclear?" *Foreign Affairs* 98, no. 1 (January–February 2019).

6 U.S. nuclear weapons modernization and the impact on the nuclear nonproliferation regime

Aiden Warren

Introduction

This chapter examines the tensions between U.S. nuclear force modernization and the global nonproliferation regime, in the context of weakening U.S. arms control/disarmament leadership and an increasingly challenging international security environment. U.S. nuclear forces, overseen and operated by the Air Force and Navy, are entering into a new era that will encompass the modernization of warheads, bombs, and delivery systems. Many of these land- air-, and sea-based systems which comprise the so-called nuclear triad, came into service during the Cold War and are now approaching the conclusive stage of their lifecycle over the next two decades. This process will see the phasing out and replacement of ballistic missiles, submarines, bombers, fighters, and air-launched cruise missiles operating with newer systems, while creating new nuclear warheads and improving facilities that produce and uphold nuclear weapons.[1] While some modernization efforts are already underway, there has been extensive debate surrounding the massive investment this will entail, the extent to which new 'flexible' modes will impact on broader global security and the strategic calculi of other states, and as this chapter will discuss, the ramifications such modernization will have on the United States in meeting its Treaty on the Nonproliferation of Nuclear Weapons (NPT) obligations.

At the beginning of his tenure in office, President Obama boldly proclaimed, "America's commitment to seek the peace and security of a world without nuclear weapons." While the tone was positive and embraced by many in the international community, his administration proceeded to initiate a plan to modernize the triad and eliminate fewer nuclear weapons than that of any other post-Cold War president.[2] In fast forwarding to Obama's successor, President Trump declared in early 2017 that he would seek to "greatly strengthen and expand [U.S.] nuclear capability," and in his administration's *Nuclear Posture Review* of February 2018, proposed two further nuclear options for the United States. The first pertained to the development of a low-yield warhead for the existing stockpile of Trident SLBMs (submarine-launched ballistic missiles). The second was a sea-launched cruise missile that would use an existing warhead. Both of these options would add to the available range of low-yield nuclear options in the U.S. suite, complementing the bomber-deliverable gravity bombs and air-launched cruise missiles. In the document, the administration articulated these applications as provisioning an essential "hedge against future nuclear 'break out' scenarios."[3] Clearly such potential capability advancements have spurred concerns relating to the security scenarios that may arise should they come to fruition, as well as the impact such U.S. modernization may have on the NPT, and the broader NPT regime.

In unpacking and assessing these debates, the chapter will commence by providing an overview of the NPT, including the recent strains it has been facing in the context of the NPT review conference process. The chapter will then proceed to assess and elucidate the role that the U.S. has played in driving the NPT and regime through its leadership and political will, albeit this has wavered at various intervals since its inception in 1968. Lastly, the chapter will look specifically at U.S. nuclear modernization and examine the tensions between U.S. modernization programs and the divisions this is producing within the nuclear NPT. Given the growing frustrations of non-nuclear weapon states over the slow pace of disarmament and a deteriorating U.S.-Russian relationship that threatens existing arms control treaties, the chapter examines the role that nuclear modernization programs play in exacerbating these fractures.

The NPT and the NPT regime

Well before the destruction of Hiroshima and Nagasaki signalled the beginning of the nuclear age, many Manhattan project scientists were intensely concerned about the challenges the atomic bomb would pose. Reflecting the fears of these scientists, the Franck Report of June 1945 signified that since an uninterrupted U.S. monopoly would be almost impossible to preserve, then the control or even the eventual elimination of nuclear weapons should best be realized through international agreements. Approximately two months after Hiroshima, President Harry Truman told Congress in October 1945: "The hope of civilization lies in international arrangements looking, if possible, to the renunciation of the use and development of the atomic bomb." Those sentiments lay behind U.S. proposals made to the United Nations in June 1946. Named the "Baruch Plan" after its chief American negotiator Bernard Baruch, a financier and friend of the president, the fundamental objective was to prevent the further spread of nuclear weapons by securing atomic technology and materials under the control of the newly created United Nations Atomic Energy Commission.

Unfortunately, the plan did not fully take into account the political and security realities that dominated the burgeoning Cold War. Established "to deal with the problems raised by the discovery of atomic energy," the Commission sought unsuccessfully to find an international agreement on control and usage. Most members of the United Nations wanted to promote the peaceful uses of atomic energy without disseminating its military technology and, thereby, preventing the proliferation of atomic weapons. After the failure of the Baruch Plan and the development of hydrogen bombs in the early 1950s, the Disarmament Commission replaced the UN Atomic Energy Commission in carrying on the search for means of halting the spread of nuclear weapons.[4] UN discussions to stem nuclear weapons proliferation continued to be productive, with the UN General Assembly unanimously approving Resolution 1665. Based on an earlier Irish draft resolution, the proposal called on negotiations in which states who already had nuclear weapons would: "refrain from relinquishing control" of them to others; "refrain from transmitting information for their manufacture to states not possessing" them; and states without nuclear weapons would agree not to receive or manufacture them. It was these sentiments that would form the basis of the 1968 Treaty on the Nonproliferation of Nuclear Weapons (NPT).[5]

Coming into effect in 1970, the NPT has today been signed and ratified by 191 different members. The treaty encompasses four main pillars: nonproliferation of nuclear weapons, nuclear weapon disarmament, peaceful use of nuclear energy, and ending nuclear arms competition. The main article of the treaty pertaining to the requirements of nuclear disarmament and nuclear competition is Article VI, which states that: "Each of the Parties to the Treaty undertakes to pursue negotiations in good faith on effective measures relating to cessation of the nuclear arms race at an early date and to nuclear disarmament, and on a treaty on general and complete disarmament under strict and effective international control."[6] Despite the fact that nearly half of the world's Nuclear Weapons States (NWS) are excluded from the treaty's provisions (North Korea, Israel, India and Pakistan), "the five nuclear weapon states party to the NPT – possess more than 98 percent of the world's nuclear weapons" and thus, the actions of these states, particularly the United States, remain paramount to the treaty's future and ultimately, its survival.[7] As stated by former Australian Foreign Minister, Gareth Evans, "without the NPT there would be considerably more nuclear-armed states than the nine clear-cut cases we have today. The well-remembered prophecy of the U.S. government in 1963, that the following two to three decades would see the emergence of 15 to 25 such states, would have been much closer to reality."[8]

In 1975, the 91 states-parties to the NPT met for the treaty's first review conference and agreed to hold future conferences to review the implementation of the treaty every five years.[9] Subsequent conferences were held in 1980, 1985 and 1990, and at the 1995 NPT Review and Extension Conference (NPTREC) the decision was made to extend the treaty indefinitely. On May 11, 1995, the Conference decided, without a vote that, "as a majority exists among States party to the Treaty for its indefinite extension, in accordance with Article X, paragraph 2, the Treaty shall continue in force indefinitely." Coinciding with this decision, the "Strengthening the review process for the Treaty" and "Principles and objectives for nuclear nonproliferation and disarmament" were also adopted. The overall decision on the extension was favored over rolling extensions of 25 years, in which there would be either one extension of a fixed period or no extension at all. The decision in 1995 to extend the treaty indefinitely still carries today significant weight and with this, greater challenges and obligations.[10] Not surprising, the seminal 1995 Conference emphasized the implementation of the provisions on disarmament (Article VI) and on safeguards and peaceful uses of nuclear energy (Articles III and IV).

In the context of Article VI, there was a decisive convergence of views between the developing and developed non-nuclear-weapon states (NNWS) on the need for the nuclear-weapon states (NWS) to proceed more promptly towards the ultimate goal of nuclear disarmament. The call has been reaffirmed at all NPT Review conferences and appeared to yield substantive agreement across all states, particularly at the 2000 conference which resulted in all NPT states agreeing to the "unequivocal undertaking by the nuclear-weapon states to accomplish the total elimination of their nuclear arsenals."[11] Similarly, at the 2010 NPT Review Conference (RevCon), the final document reaffirmed that "the nuclear-weapon states commit to accelerate concrete progress on the steps leading to nuclear disarmament . . . in a way that promotes international stability, peace and undiminished and increased security,"[12] as well as diminishing "the role and significance of nuclear weapons in all military and security concepts, doctrines and policies."[13]

Notwithstanding the positive trajectory of the NPT Review Conference of 2010, the 2015 version illustrated that the dichotomy between the NWS and NNWS had widened. At the official closing meeting of the month-long ninth Review Conference it became concerningly evident that the parties would not reach consensus agreement on a substantive final declaration. Despite the flurry of activity at the conference's conclusion, the inability to produce a consensus outcome document was attributed to what Ambassador Feroukhi identified as distinct gaps pertaining to measures towards nuclear disarmament, humanitarian aspects of nuclear weapons use, and reporting by the recognized nuclear-weapon states. Notably, the rejected draft outcome document generally sidestepped each of these points. One exception was a modest directive that would have necessitated nuclear-weapon states to submit reports on disarmament action at the 2017 and 2019 preparatory committees. The chasm between the nuclear weapon states has since remained, and with questions marks over the leadership of key state players, the future of the 2020 Review Conference and the NPT has been a topic of concern among analysts and policymakers.[14]

The continuous fragmentation of the NPT membership has the potential to negatively impact other goals, including the further entry into force of comprehensive safeguard agreements and additional protocols, ratification of the Comprehensive Test Ban Treaty, and negotiation of the Fissile Material Cut-Off Treaty, to name a few.[15] Some have even gone as far as arguing that the lackluster efforts, in particular by NWS, has emanated from and contributed to a readjustment of nuclear weapons in their respective strategic calculi, where deterrence and responses to the intensification of power politics have seen some states vie for greater 'flexibility'. Indeed, as the core theme of this book illustrates, there has been extensive investment in so-called modernization with rhetoric and planning toward more strategic, more targeted, and more usable nuclear weapons.[16]

While some delegates and policymakers earnestly attempt to keep the NPT buoyant and alive, there appears to be a broader disdain and very little understanding of the stresses the NPT is actually under. The silent atmosphere in the Assembly Hall at the UN complex during a PrepCom meeting in 2018 was described by one representative as "the quiet in a crowded room into which a hand grenade has just been tossed."[17] This somewhat embellished comparison of course was alluding to the significant issues – mentioned above – that have in recent times come to the fore in the international discourse and to which constitute a major challenge to the continued legitimacy of the treaty. The deterioration of the strategic relationship between the United States and Russia (the two leading nuclear weapon states possessing over 93 percent of the global arsenal), in conjunction with increasingly truculent statements and accusations of treaty violations, pose significant challenges. Further, the division within the NPT membership over the appropriate pathway forward for addressing the nuclear disarmament commitments under Article VI of the NPT – with two thirds supporting the new nuclear weapon prohibition treaty while a dissenting minority of the nuclear-armed states and their allies reject it – has also exacerbated tensions. Additionally, the decision of the Trump administration to withdraw from the 2015 nuclear deal with Iran (Joint Comprehensive Plan of Action; JCPOA) does not bode well for the viability of that agreement, nor other future arms control fora.[18] Of course, the multi-billion dollar nuclear force modernization programs underway in nuclear weapon possessing states illustrate the duress to which the treaty and broader regime will continue to encounter. As stated by Paul Meyer, "The NPT will celebrate the 50th anniversary of its entry into force at the 2020 Review Conference. On the basis of the present PrepCom and the state of the current geopolitical environment, it is an open question whether that Review Conference will be an occasion for celebration or mourning."[19]

U.S. leadership and impact on the NPT regime

While there are substantially fewer nuclear weapons today than during the Cold War era, the nuclear security threat remains prominent. In fact, it can be argued that the overall risks of nuclear war have expanded – wherein more states in more unstable regions have attained such weapons, terrorists continue to pursue them, and the command and control systems in even the most sophisticated nuclear-armed states remain susceptible to not only system and human error but, increasingly, to cyber-attack.[20] The failure of existing nuclear-armed states to disarm, the inability to impede new states obtaining nuclear weapons (Pakistan, India, North Korea), the potential of non-state actors gaining access to such weapons or radioactive material, and the proliferation potential of the expansion of nuclear energy all present serious security challenges. Additionally, in contradiction to the requirements of the treaty, partied NWS are no longer looking to maintain the substantial progress made towards total disarmament, but are instead working to expand and enhance their overall nuclear capabilities and thus "prolong the nuclear era indefinitely."[21] As such, the role, efforts and policy initiatives (or lack of) of the U.S. can and will have a decisive impact on the extent to which disarmament and nonproliferation efforts are strengthened.

Since the end of the Cold War, the U.S. has for the main extent played a driving role in the global nonproliferation regime: its leadership has been critical to the success or failure of virtually every nonproliferation and disarmament initiative that has been launched, including within the NPT review process. The election of Barack Obama as the 44th President of the United States promised significant changes in U.S. nuclear policy and priorities. During the 2008 Presidential campaign, Obama pledged to set a new direction in U.S. nuclear weapons policy and show the world that America believed in its existing commitment under the Nuclear Nonproliferation Treaty (NPT), and would therefore work toward the goal of eliminating all nuclear weapons.[22] In the President's pledge, he vowed to "take concrete steps towards a world without nuclear weapons" and to "put an end to Cold War thinking" by reducing "the role of nuclear weapons in our national security strategy, and urg[ing] others to do the same."[23]

Over the last 5 years, however, it is evident that the United States has been moving *away* from these pronouncements. Aside from slowing down its nuclear stockpile reductions during its second term in office, the Obama administration embarked on an overhaul of its entire nuclear weapons enterprise, encompassing the development of new weapons delivery systems and modernizing its enduring nuclear warhead types and nuclear weapons production facilities in a program that scholars estimate could cost more than a trillion dollars.[24] The modernized B61-12 warhead, for example, will be able to strike targets more accurately with a smaller explosive yield and reduce the radioactive fallout from a nuclear attack, and could thus become more militarily attractive and potentially increase the likelihood of use. Other modifications under consideration, such as interoperable warheads that could be used on land- and sea-based ballistic missiles, "could significantly alter the structure of the nuclear warheads" and require nuclear test explosions, and thus potentially introduce new uncertainties into the global nuclear order.[25] Simply put, work is under way to design new weapons to replace the current ones that will – of most concern – introduce new military capabilities to the weapons system.

Given the disarmament commitments articulated in the 2000 Final Document and the 2010 NPT Action Plan, the U.S. modernization program directly undermines these commitments. Not surprisingly, such backtracking and the overall failure of NWS to live up to their commitments has spurred tensions from those in civil society – and has in many instances inspired the humanitarian impacts movement (which held major conferences hosted by Norway, Mexico and Austria) and seen an expanding number of NNWS drive their own initiatives that could lead to faster pathways to disarmament such as the ban treaty. Yet, with the deterioration of bilateral relations between Moscow and Washington impeding any new arms control agreements, no multilateral disarmament negotiations in sight, and a clear preference towards nuclear modernization, there are substantive reasons to question the long-term prospects of the nonproliferation regime. In recent years a number of writers have developed sophisticated analyses of the development and purposes of the nonproliferation regime and the NPT;[26] analysis that is also reflected in important reports from a series of international commissions.[27] Whilst diverse, these works exhibit important common themes: they assert the significance of the Treaty for international security; examine a number of weaknesses and contradictions in the regime, notably a division between the nuclear weapons states (NWS) and non-nuclear weapon states (NNWS) over disarmament; and the importance of disarmament as an underpinning structure for nonproliferation. They all make this latter argument as a *strategic* one, even though it is also a widely held normative interpretation of the NPT regime.

However, these studies and reports all predate the emergence of the nuclear modernization agenda after the New START (Strategic Arms Reduction Treaty) was signed, and do not engage in detail with nuclear strategic postures and weapon systems. Bukovansky and colleagues in particular discuss the nonproliferation regime in relation to the "special responsibilities" of the Permanent 5 Members of the UN Security Council (P5) and especially the U.S. for international security. They identify the clashing interpretations of Article VI (in which the NNWS expected the NWS to rapidly disarm but the U.S. believed that the NPT "would do little to constrain U.S. actions in the nuclear field") as a major problem in the regime. Despite citing the 1980 demand by the Non-Aligned Movement for a "cessation of the qualitative improvement and development of nuclear weapons systems" and President Obama's renewed declaration of support for disarmament after 2008, they do not connect the issue to U.S. strategic policy or postures, or cite the commitment to nuclear modernization that resulted from negotiations with the Senate to pass New START. In short, this literature remains somewhat disconnected from questions of nuclear strategy and has yet to engage the modernization problem.

Much nuclear strategic writing has tended to be skeptical of the disarmament agenda and downplay the NPT. Likewise, O'Neil notably contests this interpretation of the regime and recommends efforts at managing a multipolar system of deterrence, which is an important reference point for understanding the pro-modernization viewpoint.[28] The International Institute of Strategic Studies, however, has published a series of papers exploring the strategic complexities of disarmament and setting out the ideal mix of approaches and strategies that can support a transition to a world with few or no nuclear weapons.[29] Setting up such a transition path was also a goal of the International Commission on Nuclear Nonproliferation and Disarmament (ICNND). Other studies[30] have critiqued the internal logic of deterrence and set out strategic paths towards zero, inspired by a range of earlier critiques from the Cold War.[31] While they all predate the modernization process, they can inform us on potential strategic policy options that are compatible with U.S. obligations under Article VI of the NPT.

Notwithstanding some analysts and writers from the think tank and media domains who have expressed their concerns about modernization, there is a lack of substantive analyses. Most have correctly argued that without some form of defined limitations on the pace and scope of nuclear modernization, the goals of deep cuts in, and eventual elimination of nuclear weapons, will remain elusive, as the continued reaffirmation of the value of nuclear weapons threatens to extend the nuclear era indefinitely.[32] Others also contend more vigorously that what we are seeing today lies somewhere between parallel efforts to refurbish Cold War arsenals and the emergence of a new arms competition fueled by enhancements to existing weapons or production of new or significantly modified types.[33] Much of the writing has focused on Washington's upgrading of the B61-4 bomb, a version that would equip the device with a tail assembly, turning it into a precision-guided stand-off-weapon.[34] The Obama administration advocated the merging of four old B61 models into a single version (B61 Model 12) that greatly reduced the range of destructive power. It would have a "dial-a-yield" feature where the lowest setting would only be 2 percent as powerful as the bomb dropped on Hiroshima in 1945. Initially, the plan seemed plausible, until attention fell on the bomb's new tail section and steerable fins. The Federation of American Scientists (FAS) argued that the high accuracy and low destructive settings mean that military commanders could press to use the bomb in an attack, knowing the radioactive fallout and collateral damage would be limited.[35]

The United States is not alone in this modernization drive. That said, while understanding that nuclear modernization is occurring among a group of nuclear states, there is a clear rationale for focusing on what the U.S. can do to progress arms control and disarmament in an environment of declining trust among the great powers, particularly in considering its strategic weight and the leadership role it has played on disarmament issues. We are still in a window where *what* U.S. experts and practitioners think and do significantly matters to the future of the nonproliferation regime. Global power shifts have meant that the window is starting to close and remains tenuously open for U.S. nonproliferation and disarmament leadership. However, the potential to make changes to its nuclear policy that would be reassuring to adversaries and stabilize strategic relationships, and which could then flow into renewed momentum on co-operative disarmament and improving legal, normative and strategic dynamics to this end looks to be waning. If nuclear force modernization programs continue apace without a renewed focus on arms control and disarmament, the future prospects of the NPT – which is almost universally acknowledged to be the cornerstone of the nuclear nonproliferation regime and one of the most important pillars of international security – are bleak. Further, nuclear modernization, particularly to the level to which the Trump administration has espoused in its *Nuclear Posture Review* of 2018, looks to open caveats in which the United States is seemingly looking for ways to "let go" of its commitment to the NPT and leadership role in the broader nonproliferation regime.

U.S. nuclear modernization

Recent substantial investments by nuclear weapon possessor states in the upkeep and modernization of their nuclear postures indicate a return of the nuclear factor in international politics – where deterrence appears to be taking precedence over nuclear arms control and disarmament, and the ultimate global goal of nuclear abolition.[36] Indeed, "all nations with nuclear weapons are modernizing their arsenals, delivery systems, and related infrastructure."[37] Nuclear modernization is thus becoming increasingly omnipresent across all NWS. Eric Schlosser's special report *The Growing Dangers of the New Nuclear-Arms Race* (2018), provides a useful overview of all the current NWS modernization programs. North Korea has most likely developed a hydrogen bomb, and its Hwasong-15 missiles may be large enough to transport not only a warhead but also decoys, chaff, and other countermeasures that would thwart America's ground-based midcourse defense anti-ballistic-missile system. India recently commissioned its second ballistic-missile submarine, launched an Agni-5 ballistic missile that can strike targets throughout Pakistan and China, and tested nuclear-capable BrahMos and Nirbhay cruise missiles.

Pakistan now has the world's fastest-growing nuclear stockpile, including low-yield warheads on Hatf-9 missiles for use against Indian troops and armored vehicles. Israel is expanding the range of its Jericho III ballistic missiles and deploying cruise missiles with nuclear weapons on submarines. France and the United Kingdom are developing replacements for their Vanguard and Triomphant ballistic-missile submarines. China is about to introduce Dongfeng-41 ballistic missiles that will be mounted on trucks, loaded with up to ten nuclear warheads, and capable of reaching anywhere in the United States. Russia is building a wide range of new missiles, bombers, and submarines that will carry nuclear weapons. The R-28 Sarmat missile, nicknamed Satan-2, will carry up to sixteen nuclear warheads – more than enough for a single missile to destroy every American city with a population larger than a million people. Russia plans to build 40 to 50 of the Satan-2s.[38]

In the context of the United States, the specific logistical and technical aspects of its nuclear modernization plans are categorized as follows: *(1) the refurbishment of nuclear warheads; (2) modernization of strategic delivery systems; (3) refinement of the command and control systems; (4) adjustment of the production complex; (5) implementation of the Nuclear Force Improvement Program.*

(1) Insofar as the *refurbishment of nuclear warheads*, notwithstanding the fact that the U.S. stockpile of nuclear warheads and bombs is perpetually refurbished via the National Nuclear Security Administration's (NNSA) Life Extension Program (LEP) and have been deemed as being to be safe and reliable, the modernization drive is still being pursued. In this regard, the NNSA is presently undertaking a contentious and costly plan to unify the current amount of nuclear warhead types from 10 down to 5. Known as the "3+2" strategy, the five LEPs connected with this approach are projected to cost in the vicinity of $60 billion.

(2) In terms of *modernizing strategic delivery systems*, this will involve the comprehensive rebuilds of the Minuteman III ICBM and Trident II SLBM. The operational tenure of the Navy's 14 Trident Ohio-class ballistic missile submarines are also being prolonged. Furthermore, a new class of submarine (the Columbia) which will take over from the Ohio-class ballistic missile submarines, is undergoing development and is projected to cost approximately $128 billion to construct, according to the Defense Department. The relatively new B-2 strategic bomber, is also being enhanced, as is the B-52H bomber. The Air Force is also preparing for a new strategic bomber, the B-21, as well as a new nuclear-capable cruise missile, known as the Long-Range

Standoff Weapon (LRSO) to exchange for the current Air-Launched Cruise Missile (ALCM).[39]

(3) The *command and control systems* are also set to undergo a revitalization. It is the Defense Department that sustains the command, control, communications, and early-warning systems which enable those in the field to communicate with nuclear forces, signal commands that define their use, and identify and/or mitigate external attacks. While the department estimates that it will spend $40.5 billion on such activities between FY 2017 and FY 2026, it is most likely an understated figure as the Pentagon is still orchestrating its modernization plans for such systems. It also needs to be noted that the Trump 2018 NPR has demanded and emboldened emphasis on sustaining and upgrading command and control capabilities, which will also impact the figure.[40]

(4) *Modernizing the production complex* is also well underway, incorporating the planning and creation of new facilities, as well as a projected budget. Here, the FY 2019 NNSA budget request is looking to attain $703 million for the Uranium Processing Facility (UPF) at Oak Ridge, Tennessee. According to an autonomous research project undertaken by the Corps of Engineers, the entire construction cost for UPF is projected to be in the range of $6.5–7.5 billion; although some evaluations put the cost at the astonishing figure of $11 billion. The disparity is quite marked when considering that NNSA has promised to conclude the construction by 2025 for $6.5 billion.[41]

(5) The last of the categories pertains to the *nuclear force improvement program*. Based on various disclosures of professional and ethical slips and average assurance in the U.S. nuclear force, Defense Secretary Chuck Hagel proclaimed in November 2014 that the department would undertake measures to address the setbacks. These comprised of changing the conduct of inspections to reduce the burden on airmen and sailors, "eliminating micromanagement of nuclear personnel seen as overtaxed by excessive bureaucratic and administrative requirements, and elevating the head of Air Force Global Strike Command, which oversees the Air Force's nuclear forces, from a three- to a four-star rank."[42]

Additionally, under the mask of a more palatable warhead life-extension program, the U.S. military has inaudibly emboldened and increased the devastating power of the highest volume warhead contained in the U.S. nuclear arsenal (the W76, deployed on the Navy's ballistic missile submarines). This enhancement in destructive power means that all U.S. sea-based warheads now have the capacity to destroy subterranean sites including Russian missile silos, an ability formerly saved specifically for the highest-yield warheads in the U.S. stockpile. Such development has occurred away from the consumption of most government officials and policy makers, who have been more focused on lowering nuclear warhead numbers. As such, what has been happening is that the nuclear arsenal has been transitioning toward a force capacity that encompasses explicit features specifically aimed at undertaking surprise attacks against Russia and for executing "successful" nuclear wars. In this regard, the paradoxical approach clearly embodies, on one hand, a marked increase in lethality and firepower, contrasted with a reduction in the number of weapons possessed by U.S. and Russian forces. Of course, this could engender a significant increase in the susceptibility of Russian nuclear forces to a U.S. first strike, and with that, a form of response. According to some analysts, the contrast of arms reductions, combined with an increase in U.S. nuclear capacity means the United Stated could potentially terminate "all of Russia's ICBM silos using only about 20 percent of the warheads deployed on U.S. land- and sea-based ballistic missiles." Additionally, the prospect of super-fuze technology could make it feasible for every SLBM and ICBM warhead in the U.S. suite to execute the subterranean assignments that were originally intended to be aligned to MX Peacekeeper ICBM warheads.[43]

Overall, the W76 advancement signifies a marked adjustment in which the U.S. hard-target kill capability has shifted from land-based to sea-based ballistic missiles. In moving this capability to submarines that can position themselves much closer to launch missiles on their targets than land-based missiles, the U.S. military has secured a markedly improved capability to execute a covert first strike against Russian ICBM silos. An increased sea-based offensive and defensive capacity, however, could potentially spur the chances of conflict and ultimately nuclear war with either Russia or China. With Russian silos more susceptible to W76–1/Mk4A warheads, the overall expanding capability of the U.S. forward-deployed sea-based nuclear missiles could engender serious concerns in the minds of Russian military-political leaders pertaining to U.S. strategic intentions – particularly in light of expanding U.S. cyber, innovative conventional, and missile defense capabilities – and almost certainly intensifying suspicion and worst-case planning calculations in Russia.[44]

Not surprisingly, these developments are likely to have serious implications for global disarmament efforts and are of particular concern given Washington's capacity to act as a circuit breaker in the global nuclear order. In looking at the disarmament commitments articulated in the 2000 Final Document and the 2010 NPT Action Plan, U.S. modernization plans, as conveyed in the above, directly undermine these commitments, and run counter to Obama's 2009 promises laid out at Prague. Further, with the deterioration of bilateral relations between Moscow and Washington impeding any new arms control agreements (not to mention actual departures), no multilateral disarmament negotiations in sight, and a clear preference towards nuclear modernization by both states, there are good reasons to question the long-term survival prospects of the nonproliferation regime.

The Trump administration

Of course, the amplified declaratory statements and policy releases of the Trump administration appear to have only widened the gap between the United States modernization drive and meeting its obligations under the NPT. The Trump *Nuclear Posture Review*, for instance, illustrates some significant shifts away from the Obama review of 2010. The most noteworthy adjustment appears to be a move away from seeking to decrease the number of U.S. nuclear weapons and their role in U.S. military strategy. In its place, the Trump review has embraced a more assertive tone and has presented a posture that clearly seeks to embolden the reliance on nuclear weapons, including plans to create new nuclear weapons while also modifying and enhancing others. Additionally, the report has seceded from the goal of establishing nuclear weapons' sole purpose as deterring nuclear attacks, and in more forthright fashion emphasizes a role for nuclear weapons in deterring "non-nuclear strategic attacks," and even cyber-attacks. As a means to achieve this, the review stipulates that "the United States will enhance the flexibility and range of its tailored deterrence options. . . . Expanding flexible U.S. nuclear options now, to include low-yield options, is important for the preservation of credible deterrence against regional aggression."[45]

Over the long term, the Trump NPR asserts that the United States will "pursue a nuclear-armed" submarine-launched cruise missile so as to "provide a needed nonstrategic regional presence, an assured response capability, and [in light of] Russia's continuing . . . violation of the Intermediate-Range Nuclear Forces Treaty (INF Treaty)," a response that it believes is acquiescent with the treaty. By pursuing this new type of missile, the document argues that the United States "will immediately begin efforts to restore this capability by initiating a requirements study leading to an Analysis of Alternatives . . . for the rapid development of a modern [submarine-launched cruise missile]." Additionally, the writers of the NPR argue that the "U.S. pursuit of a submarine-launched cruise missile may provide the necessary incentive for Russia to negotiate seriously a reduction of its non-strategic nuclear weapons, just as the prior Western deployment of intermediate-range nuclear forces in Europe led to the 1987 [Intermediate-Range Nuclear Forces] Treaty."[46] Further, the NPR declares that the nuclear arsenal will provide a more varied set of features that will enhance the capacity to refine "deterrence and assurance; expand the range of credible U.S. options for responding to nuclear or non-nuclear strategic attack; and, enhance deterrence by signaling to potential adversaries that their concepts of coercive, limited nuclear escalation offer no exploitable advantage."[47]

Aside from the security implications that the modernization drive will yield, the scope of such an undertaking, encompassing all elements of the nuclear arsenal and the production complex within it, is significant to say the least. Notwithstanding frequent concerns expressed pertaining to the exorbitant cost of the modernization program in its current formations, the NPR has not highlighted nor addressed this serious economic challenge. Whether Congress will endorse the funding of the costly programs in lieu of constructing modest and cost-efficient life-extended versions of existing designs, remains to be seen. Additionally, significantly restructuring warheads would challenge the pledge of the 2010 U.S. *Nuclear Posture Review* Report that the United States "will not develop new nuclear warheads" but instead consider the "full range" of life-extension program options, including "refurbishment of existing warheads, reuse of nuclear components from different warheads, and replacement of nuclear components."[48] Of course, this statement was projected to mitigate the recommencement of nuclear explosive testing and observe the 1996 Comprehensive Nuclear Test Ban Treaty. Further, the Trump NPR signified that any life-extension programs would "use only nuclear components based on previously tested designs and will not support . . . new military capabilities." In this regard, adherence rests on the definitional basis of what constitutes "new" military capabilities, particularly when considered in the context of new or improved dimensions outside the nuclear explosive suite that may embolden a weapon's military proficiencies. In relation to the Trump administration, however, it is evident that a strategy encompassing improvement of nuclear weapons' accuracy and agility via lower yield of modified warheads will be par for the course and an integral plank in its overall nuclear strategy.

The demise of the INF Treaty

Further evidence of the Trump administration's drive to redefine and readjust the United States' nuclear weapons position is witnessed in the withdrawal from the INF Treaty. In October 2018, President Trump loosely detailed at a Nevada rally that he would withdraw the United States from the Intermediate-Range Nuclear Forces Treaty (INF) with Russia. The pivotal 1987 agreement prohibits both the U.S. and Russia from possessing or manufacturing ground-launched cruise missiles with a range between 480 to 5,470 kilometers. For several years, the U.S has accused Russia of breaching this deal through the development of a variety of missiles, including "the Novator 9M729 – known to NATO as the SSC-8." In his announcement, Trump stated that the U.S. was "not going to let them violate a nuclear agreement and go out and do weapons and we're not allowed to . . . if Russia's doing it and if China's doing it" then "we are going to develop the weapons."[49]

The main concern with the INF departure is that the Administration did not do anything substantive to persuade or coerce Russia back into compliance, which should have been the defining goal. In fact, it virtually eradicated legal and political pressure on Russia and essentially gave the Kremlin carte blanche to develop and deploy INF systems in greater numbers and without any treaty limitations. To some extent, these developments are hardly surprising given the deterioration of bilateral relations over the last 5–6 years. While Russia under Putin has regularly contravened international agreements, the Trump administration has also undertaken a pattern of withdrawing from such agreements. The President has removed America from more than a dozen international commitments, pointing to an imperative to reassert America's international control by "letting go," and undermining the very post-World War II liberal international order that the United States defined and orchestrated. Given the INF Treaty's crucial stabilizing contribution to Europe's security, America's NATO allies once again questioned the commitment of Trump to the organization itself and broader EU security. That said, according to Kimball and Reif, if NATO member states really wanted to sustain a key arms control treaty that has contributed to their security for more than two decades, they needed to be more assertive and active in insisting that the United States and Russia pursue diplomatic options at an earlier date.[50]

Clearly, Trump's preparedness to end reciprocally beneficial agreements so as to placate his domestic base does not augur well for global stability. On November 4, 2018, Secretary of State Mike Pompeo declared Russia was in material breach of the landmark 1987 Intermediate-Range Nuclear Forces (INF) Treaty and that the United States planned to suspend U.S. requirements under the treaty in 60-days unless Russia returned to compliance. While NATO foreign ministers collectively agree, declaring "that Russia has developed and fielded a missile system, the 9M729, which violates the INF Treaty [and] it is now up to Russia to preserve the INF Treaty," the Trump administration's thin 60 day demand and "hope" that Russia would "change course," came across as a concession that the INF Treaty was for all intents and purposes dead.[51] On February 2, 2019, the United States provided a six-month notice of withdrawal from the Intermediate-Range Nuclear Forces (INF) Treaty due to the Russian Federation's continuing violation of the treaty. On August 2, 2019, pursuant to Article XV of the treaty, the United States formally withdrew from the treaty "because Russia failed to return to full and verified compliance through the destruction of its noncompliant missile system – the SSC-8 or 9M729 ground-launched, intermediate-range cruise missile."[52]

The non-renewal of New START

The United States' INF departure can be considered a precursor to the Trump administration's desire to terminate yet another Obama signature agreement. While bilateral relations have been significantly strained since 2012, the U.S.-Russian arms control architecture in the context of New START has remained one of the positive lights of the relationship. Ratified in 2011, the Treaty places a limit on the number of deployed strategic warheads to a maximum of 1,550 in both states, a target each met early in 2018, and which is extensively below the tens of thousands of weapons that peaked in 1986. While New START expires in February 2021, it can be extended by up to five years by agreement by the two Presidents without necessitating additional actions by the Congress or the Duma. Should the treaty not be extended for the period after 2021, then there will be no legally binding limitations for the first time since 1972 on the two states that possess 93 percent of the world's nuclear weapons. Most concerning is that without substantive impediments there is the possibility of a spiral into an arms race, which would add billions in further costs to the U.S. nuclear modernization drive.

Indeed, without any substantive limitations in the form of New START, both states would most likely move apace to increase the number of weapons deployed. As such, an extension until February 2026 would maintain its crucial security advantages – not only in terms of numerical limits, but also the reciprocated pellucidity that the treaty's verification mechanisms (encompassing data exchanges, notifications, and inspections) facilitate. More importantly, an extension would also enable more time for a political and diplomatic solution, as well as providing an opportunity to deliberate and conceivably limit various new systems under development by Russia, and provide the foundation for dialogue that would contribute to decreasing both states' nuclear arsenals. Additionally, while many analysts and commentators are worried about what the Trump administration might relinquish or trade away with Putin, extending New START could foster a constructive environment conducive to lowering tensions in bilateral relations without offering imprudent or unreasonable concessions to Russia. Lastly, in adapting to Trump's penchant for branding, and in a somewhat unorthodox approach, an extension of New START could even be rebadged as a personal triumph for President Trump, instead of an extension of an Obama foreign policy/diplomatic success.[53]

Despite many Washington DC analysts having queried the link between the INF Treaty departure/non-New START renewal and the potential to trigger the modernization drive, there are others who are deeply concerned that such developments may spur a modernization arms race. In this regard, they argue, the demise of the only U.S.-Russia arms control pact limiting deployed nuclear weapons would make it more difficult for both states to assess the other's intentions, giving both inducements to enlarge and improve their arsenals.[54] Indeed, arms control agreements remain a significant tool for ascertaining parity between Russia and the U.S., and provide a central point for discourse on a variety of matters relating to military balance.[55]

The New START Treaty achieved an important milestone in February 2018 when both states proclaimed their respective reductions of their strategic forces completed. However, with the prospects for extending New START or for negotiating a new treaty that would replace it being unclear – as well as the demise of the Intermediate-Range Nuclear Forces (INF) Treaty – it appears that both parties are preparing for greater agility. Russia clearly wants to invest in a range of new systems and capabilities, perceiving traditional arms control mechanisms as being unable to dependably limit strategic competition or constrain several important capabilities (from missile defense to conventional strike weapons, and latent capabilities, such as weapons in space). Further, it has also questioned the extent to which treaties, such as New START, can actually limit the number of strategic launchers and warheads.[56] Likewise, from the United States' point of view, there has been a drive by the national security/stern deterrence camp to press for greater 'flexibility', not only in response to perceived Russian overtures, but also to China's own military and strategic aspirations. What was once a somewhat uneasy bipartisan consensus during the earlier stages of the Obama administration, has now descended into a marked and intensified division that could well undermine the arms control process.[57] As stated by Senator Robert Menendez (D-NJ), ranking member of the Senate Foreign Relations Committee, at a September 18, 2018, hearing on U.S.-Russia strategic arms control:

> I also want to remind the administration that bipartisan support for nuclear modernization is tied to maintaining an arms control process that controls and seeks to reduce Russian nuclear forces, which inevitably means promoting militarily- and fiscally-responsible policies on ourselves. We are not interested in writing blank checks for a nuclear arms race with Russia. And we

> don't want to step off our current path of stability to wander again down an uncertain road filled with potentially dire consequences.[58]

Conclusion

The United States is poised to spend more than $1.7 trillion over the next 30 years on maintaining and upgrading its nuclear delivery systems (bombers, land-based missiles, and submarines) and their associated warheads and supporting infrastructure. The Trump administration's 2018 *Nuclear Posture Review* clearly specified the conditions under which the use of nuclear weapons would be considered and proposed two new, "more usable" types of low-yield nuclear weapons.[59] Proponents of modernization (and many within the U.S. nuclear establishment) have argued that such changes are *merely* refurbishments/improvements rather than wholesale redesigns, and thereby, fulfill the pledge of not producing a new nuclear weapon.[60] However, much of the recent literature has focused on the newer "agile" applications with concern. These analysts argue that while the explosive innards of the revitalized weapons may not be entirely new, given their smaller yields and better targeting such devices might be seen as *more* conceivably usable in a limited or tactical conflict – even to use first, rather than in retaliation.[61] While official statements continue to justify nuclear modernization as simply extending the service life of existing capabilities, the "Pentagon now explicitly paints the nuclear modernization as a direct response to Russia."[62]

Notwithstanding the marked reductions in the overall number of nuclear weapons since the Cold War's demise, all of the world's nine nuclear-armed states are busily modernizing their remaining nuclear forces and appear to be clearly in it for the long haul.

> None of the nuclear-armed states appears to be planning to eliminate its nuclear weapons anytime soon. Instead, all speak of the continued importance of nuclear weapons. . . . Perpetual nuclear modernization appears to undercut the promises made by the five NPT nuclear-weapon states.[63]

Under the terms of that treaty, they are required to "pursue negotiations in good faith on effective measures relating to cessation of the nuclear arms race at an early date and to nuclear disarmament."[64] Fifty years after this promise was specified, the non-nuclear-weapon states (NNWS), who in return for that commitment renounced nuclear weapons for themselves, can rightly challenge the notion of whether continued nuclear modernization in perpetuity adheres to the core principles of the NPT.[65] In the context of the U.S., while it certainly is not alone in this pursuit, it precariously remains the central driving force in the liberal international order – the very order it created and defined after 1945. But "if you squint," according to Rose Gideon, and "the closer you look, the more you see it [U.S. foreign policy] being hollowed out, with the forms and structures still in place but the substance and purpose draining away."[66] In looking at the U.S.' amplified modernization strategy, it is evident that it totally scrambles the calculus of its previous nuclear disarmament and nonproliferation efforts, and with this, is challenging and "hollowing out" the aging underpinnings of the NPT that it was so integral in creating.[67]

Notes

1 Ankit Panda, "U.S. Nuclear Weapons Modernization," *Council of Foreign Relations*, February 7, 2018, www.cfr.org/backgrounder/us-nuclear-weapons-modernization.
2 Ibid.
3 U.S. Department of Defense, Office of the Secretary of Defense, "Nuclear Posture Review 2018," February 2018, https://dod.defense.gov/News/SpecialReports/2018NuclearPostureReview.aspx.
4 Susanna Schrafstetter and Stephen Twigge, *Avoiding Armageddon: Europe, the United States, and the Struggle for Nuclear Non-proliferation, 1945–1970* (Westport, CT: Praeger, 2004), 13; Special Message to Congress on Atomic Energy, October 3, 1945, *Public Papers: Truman: 1945*; Baruch's Address to the United Nations, June 14, 1946, reproduced at *Nuclear Age Peace Foundation*, NuclearFiles.org; Elements of this chapter have been adapted Professor Siracusa's study (with Richard Dean Burns), *A Global History of the Nuclear Arms Race: Weapons, Strategy, and Politics*, 2 vols. (Santa Barbara, CA: Praeger, 2013).
5 Arms Control Association, "Timeline of the Nuclear Non-proliferation Treaty (NPT)," December 2018, www.armscontrol.org/factsheets/Timeline-of-the-Treaty-on-the-Non-Proliferation-of-Nuclear-Weapons-NPT.
6 United Nations Office for Disarmament Affairs (UNODA), "Treaty on the Non-Proliferation of Nuclear Weapons (NPT)," 1970, www.un.org/disarmament/wmd/nuclear/npt/text/.
7 Hans M. Kristensen and Robert S. Norris, "Slowing Nuclear Weapon Reductions and Endless Nuclear Weapon Modernizations: A Challenge to the NPT," *Bulletin of the Atomic Scientists* 70, no. 4 (2014).
8 Gareth Evans and Yoriko Kawaguchi, *Eliminating Nuclear Threats: A Practical Agenda for Global Policymakers* (Canberra and Tokyo: International Commission on Nuclear Non-proliferation and Disarmament, 2009), 32.
9 Arms Control Association, "Timeline of the Nuclear Non-Proliferation Treaty (NPT)."
10 Reaching Critical Will, "History of the NPT 1975–1995," www.reachingcriticalwill.org/disarmament-fora/npt/history-of-the-npt-1975-1995.
11 United Nations, "Nuclear-Weapon States Endorse Goal of 'Total Elimination' of Nuclear Arsenals as NPT Review Conference Concludes," UN Meetings Coverage and Press Releases, United Nations, New York, May 22, 2000, www.un.org/press/en/2000/20000522.dc2710.doc.html.
12 United Nations, "The 2010 Review Conference of the Parties to the Treaty on the Non-Proliferation of Nuclear Weapons (NPT)," Final Documents of the 2010 NPT Review Conference, United Nations, New York, 2010.
13 Ibid.
14 Wilfred Wan, "Why the 2015 NPT Review Conference Fell Apart," *United Nations University Centre for Policy Research*, May 28, 2015, https://cpr.unu.edu/why-the-2015-npt-review-conference-fell-apart.html.
15 Ibid.
16 Statement by Ireland, "Diplomatic Conference to Negotiate a New Legal Instrument for the Prohibition of Nuclear Weapons Leading to Their Total Elimination-Organizational Meeting," United Nations, New York, February 16, 2017, www.dfa.ie/media/dfa/alldfawebsitemedia/ourrolesandpolicies/intpriorities/womenpeaceandsecurity/NWP-TOrganisationalMeeting16Feb.pdf.

17 Paul Meyer, "Sleepwalking Towards the 2020 Review of the Nuclear Non-Proliferation Treaty," *OpenCanada.org*, May 8, www.opencanada.org/features/sleepwalking-towards-2020-review-nuclear-non-proliferation-treaty/.
18 Ibid.
19 Ibid.
20 Gareth Evans and Ramesh Thakur (eds.), *Nuclear Weapons: The State of Play* (Canberra: The Australian National University, 2013); Brian Martin, "Nuclear Winter: Science and Politics," *Science and Public Policy* 15, no. 5 (1988); Peter King, "Undermining Proliferation: Nuclear Winter and Nuclear Renunciation," The Centre for Peace and Conflict Studies, Working Paper No. 09/1, October 2009; Alan Robock, Luke Oman, Georgiy L. Stenchikov, Owen B. Toon, Charles Bardeen, and Richard P. Turco, "Climatic Consequences of Regional Nuclear Conflicts," *Atmospheric Chemistry and Physics* 7 (2007), 2003–2012.
21 Kristensen and Norris, "Slowing Nuclear Weapon Reductions and Endless Nuclear Weapon Modernizations: A Challenge to the NPT."
22 See, Aiden Warren, *The Obama Administration's Nuclear Weapon Strategy: The Promises of Prague* (New York: Routledge, 2014); Joseph M. Siracusa, *Nuclear Weapons: A Very Short Introduction*, 2nd edition (Oxford: Oxford University Press, 2015).
23 Barack Obama, "Remarks by President Barack Obama," Hradčany Square, Prague, Czech Republic, The White House, Office of the Press Secretary, Washington, DC, April 5, 2009.
24 Jon B. Wolfsthal, Jeffrey Lewis, and Marc Quint, "The Trillion Dollar Nuclear Triad," *James Martin Center for Non-Proliferation Studies*, January 2014; Kingston Reif, "U.S. Nuclear Modernization Programs," *Arms Control Association*, October 2016, www.armscontrol.org/factsheets/USNuclearModernization; William J. Broad and David E. Sanger, "U.S. Ramping Up Major Renewal in Nuclear Arms," *New York Times*, September 22, 2014.
25 Hans M. Kristensen, "Nuclear Weapons Modernization: A Threat to the NPT?" *Arms Control Association*, www.armscontrol.org/act/2014_05/Nuclear-Weapons-Modernization-A-Threat-to-the-NPT.
26 William Walker, "Nuclear Enlightenment and Counter-Enlightenment," *International Affairs* 83, no. 3 (2007), 431–453; William Walker, "Nuclear Order and Disorder," *International Affairs* 76, no. 4 (2000), 703–724; William Walker, *A Perpetual Menace: Nuclear Weapons and International Order* (New York: Routledge, 2011); Shampa Biswas, *Nuclear Desire: Power and The Postcolonial Nuclear Order* (Minneapolis: The University of Minnesota Press, 2014); William Potter and Guakhar Mukhatatzanova, *Nuclear Politics and the Non-Aligned Movement* (London: Routledge and IISS, 2012); Ramesh Thakur (ed.), *Nuclear Weapons and International Security: Collected Essays* (London and New York: Routledge, 2015); Jane Boulden, Ramesh Thakur, and Thomas G. Weiss (eds.), *The United Nations and Nuclear Orders* (Tokyo, New York and Paris: United Nations University Press, 2009).
27 Evans and Kawaguchi, *Eliminating Nuclear Threats: A Practical Agenda for Global Policymakers Non-proliferation.*
28 *See*, Andrew O'Neil, "Nuclear Weapons and Non-Proliferation: Is Restraint Sustainable?" *Security Challenges* 5 (2009).
29 G. Perkovich and James M. Acton, *Abolishing Nuclear Weapons*, Adelphi Paper 396 (London: IISS and Routledge, 2009); James M. Acton, *Deterrence During Disarmament*, Adelphi Paper 417 (London: IISS and Routledge, 2011); Cortwright and Väryrynen, *Towards Nuclear Zero*, Adelphi Paper 410 (London: IISS and Routledge, 2011).
30 A. Burke, 'Nuclear Reason: At the Limits of Strategy', *International Relations* 23, no. 4 (2009), 506–529; G.P. Schultz and J.P. Goodby (eds.), *The War That Must Never Be Fought: Dilemmas of Nuclear Deterrence* (Stanford: Hoover Institution Press, 2015).

31 Hans J. Morgenthau, "The Four Paradoxes of Nuclear Strategy," *American Political Science Review* 58, no. 1 (1964), 23–35; P. Green, *Deadly Logic: The Theory of Nuclear Deterrence* (Columbus, OH: Ohio State University Press, 1966).
32 Kristensen, "Nuclear Weapons Modernization: A Threat to the NPT?"; Tanya Ogilvie-White, "Great Power Responsibility and Nuclear Order," *The Non-Proliferation Review* 20, no. 1 (2013); Wolfsthal, Lewis, and Quint, "The Trillion Dollar Nuclear Triad"; Reif, "U.S. Nuclear Modernization Programs."
33 Hans M. Kristensen, "Video Shows Earth-Penetrating Capability of B61-12 Nuclear Bomb," *FAS Blog*, Federation of American Scientists, 2016; Warren, *The Obama Administration's Nuclear Weapon Strategy: The Promises of Prague*; Siracusa, *Nuclear Weapons: A Very Short Introduction.*
34 John Mecklin, "Disarm and Modernize," *Foreign Policy*, Issue 211, March–April 2015, https://foreignpolicy.com/2015/03/24/disarm-and-modernize-nuclear-weapons-warheads/; Kristensen, "Video Shows Earth-Penetrating Capability of B61-12 Nuclear Bomb."
35 William J. Broad and David E. Sanger, "As U.S. Modernizes Nuclear Weapons, 'Smaller' Leaves Some Uneasy," *New York Times*, January 11, 2016; Kevin Robinson-Avila, "Overhauling the Nation's Nuclear Arsenalandia National Labs Achieves B61 Milestone," *Albuquerque Journal*, May 18, 2014.
36 Reif, "U.S. Nuclear Modernization Programs."
37 R. Acheson (ed.), *Assuring Destruction Forever: Nuclear Weapon Modernization Around the World* (New York: Reaching Critical Will, Women's International League for Peace and Freedom, United Nations, 2012), 88.
38 Eric Schlosser, "The Growing Dangers of the New Nuclear-Arms Race," *The New Yorker*, May 24, 2018, www.newyorker.com/news/news-desk/the-growing-dangers-of-the-new-nuclear-arms-race.
39 Arms Control Association, "U.S. Nuclear Modernization Programs," December 2018, www.armscontrol.org/factsheets/USNuclearModernization.
40 Ibid.
41 Ibid.
42 Ibid.
43 Hans M. Kristensen, Matthew McKinzie, and Theodore A. Postol, "How U.S. Nuclear Force Modernization Is Undermining Strategic Stability: The Burst-Height Compensating Super-Fuze," March 1, 2017, https://thebulletin.org/2017/03/how-us-nuclear-force-modernization-is-undermining-strategic-stability-the-burst-height-compensating-super-fuze/.
44 Ibid.
45 U.S. Defense Department, Office of the Secretary of Defense, "Nuclear Posture Review 2018," February 2018, 34, https://fas.org/wp-content/uploads/media/2018-Nuclear-Posture-Review-Version-2.pdf.
46 Ibid., 35.
47 Ibid.
48 U.S. Defense Department, "Increasing Transparency in the U.S. Nuclear Weapons Stockpile," *Fact Sheet*, May 3, 2010, www.defense.gov/Portals/1/features/defenseReviews/NPR/10-05-03_Fact_Sheet_U.S._Nuclear_Transparency__FINAL_w_Date.pdf.
49 Julian Borger and Martin Pengelly, "Trump Says U.S. Will Withdraw from Nuclear Arms Treaty with Russia," October 21, 2018, www.theguardian.com/world/2018/oct/20/trump-us-nuclear-arms-treaty-russia.
50 Daryl G. Kimball and Kingston A. Reif, "U.S. INF Treaty Termination Strategy Falls Short," *Arms Control Association* 10, no. 10 (December 4, 2018), www.armscontrol.org/issue-briefs/2018-12/us-inf-treaty-termination-strategy-falls-short.
51 Matt Korda and Hans M. Kristensen, "Trump Falls on Sword for Putin's Treaty Violation," *Bulletin of Atomic Scientists*, October 29, 2018, https://thebulletin.org/2018/10/

trump-falls-on-sword-for-putins-treaty-violation/?utm_source=Twitter&utm_medium=Twitter%20Post&utm_campaign=TrumpFallsSword.

52 U.S. Department of State, "U.S. Withdrawal from the INF Treaty on August 2, 2019, Press Statement, Michael R. Pompeo, Secretary of State," August 2, 2019, www.state.gov/u-s-withdrawal-from-the-inf-treaty-on-august-2-2019/.

53 Thomas M. Countryman, Kingston A. Reif, and Daryl G. Kimball, "Trump and Putin Can Put the Brakes on a New, Potentially More Dangerous, Arms Race," *IDN InDepthNews*, August 2018, www.indepthnews.net/index.php/opinion/2046-trump-and-putin-can-put-the-brakes-on-a-new-potentially-more-dangerous-arms-race.

54 Arshad Mohammed and Jonathan Landay, "Treaty's End Would Give U.S., Russia Impetus to Make More Nukes: Study," *Reuters*, April 1, 2019, www.reuters.com/article/us-usa-russia-nuclear/treatys-end-would-give-u-s-russia-impetus-to-make-more-nukes-study-idUSKCN1RD1AI.

55 Pavel Podvig, "Nuclear Strategies in Transition Russia's Current Nuclear Modernization and Arms Control," *Journal for Peace and Nuclear Disarmament* 1, no. 2 (2018), ORCID: Iconwww.tandfonline.com/doi/full/10.1080/25751654.2018.1526629.

56 The Ministry of Foreign Affairs of the Russian Federation, "Comment by the Information and Press Department on the Latest Data regarding the Aggregate Numbers of U.S. Strategic Offensive Arms Published by the U.S. Department of State," *The Ministry of Foreign Affairs of the Russian Federation*, February 27, 2018, www.mid.ru/foreign_policy/news/-/asset_publisher/cKNonkJE02Bw/content/id/3100658; Podvig, "Nuclear Strategies in Transition Russia's Current Nuclear Modernization and Arms Control."

57 Frank A. Rose, "The End of an Era? The INF Treaty, New START, and the Future of Strategic Stability," Brookings Institution, February 12, 2019, www.brookings.edu/blog/order-from-chaos/2019/02/12/the-end-of-an-era-the-inf-treaty-new-start-and-the-future-of-strategic-stability/.

58 Robert Menendez, "Opening Remarks at Hearing on U.S.-Russia Arms Control Efforts Washington," Senate Foreign Relations Committee, September 18, 2018, www.foreign.senate.gov/press/ranking/release/menendez-opening-remarks-at-hearing-on-us-russia-arms-control-efforts-.

59 Countryman, Reif, and Kimball, "Trump and Putin Can Put the Brakes on a New, Potentially More Dangerous, Arms Race."

60 Marina Malenic, "U.S. Completes First B-61 LEP Flight Test," *Janes Defense Weekly*, July 8, 2015; *See*, Matthew Kroenig, "How to Approach Nuclear Modernization? A U.S. Response," *Bulletin of the Atomic Scientists* 71, no. 3 (2015); Matthew Kroenig, "Why U.S. Nuclear Modernization Is Necessary," *Bulletin of the Atomic Scientists* online (January 8, 2015).

61 Mecklin, "Disarm and Modernize"; Broad and Sanger, "As U.S. Modernizes Nuclear Weapons, 'Smaller' Leaves Some Uneasy."

62 "Pentagon Portrays Nuclear Modernization as a Response to Russia," *Federation of American Scientists (FAS)*, February 11, 2016.

63 Kristensen, "Nuclear Weapons Modernization: A Threat to the NPT?"

64 United Nations Office for Disarmament Affairs (UNODA), "Treaty on the Non-Proliferation of Nuclear Weapons (NPT)," United Nations, New York, 2019, www.un.org/disarmament/wmd/nuclear/npt/text/.

65 In fact, many NNWS have already done so with, as of July 31, 2019, 24 nations having ratified the Treaty on the Prohibition of Nuclear Weapons (TPNW).

66 Rose Gideon, "Letting Go," *Foreign Affairs* (March–April 2018), www.foreignaffairs.com/articles/2018-02-13/letting-go.

67 Mecklin, "Disarm and Modernize."

7 NATO nuclear modernization

Steven Pifer

With 2020 well underway, the North Atlantic Treaty Organization (NATO) is in the process of modernizing its nuclear forces, both strategic and nonstrategic. NATO nuclear modernization is taking place against the backdrop of a European security environment that has changed dramatically and for the worse since the 1990s, particularly over the past five years. Russia has made clear its dissatisfaction with the post-Cold War security order and is modernizing its nuclear and conventional forces. Questions have emerged, moreover, about Russian nuclear doctrine and its readiness to resort to nuclear weapons in a conflict.

These developments and Moscow's apparent embrace of an adversarial approach toward NATO affect the alliance's posture and modernization plans. NATO's recent summit statements have devoted increasingly detailed language to the nuclear component of the alliance's deterrence and defense posture. The United States, Britain, and France have modernized or are about to modernize their strategic nuclear forces, and other individual allies will modernize their dual-capable aircraft, the planes that could deliver U.S. B61 non-strategic nuclear bombs, which are also being modernized.

As NATO proceeds with its nuclear modernization, the alliance should consider how to enhance deterrence while also seeking to contain and manage tensions with Russia. NATO should structure its nuclear posture so as not to lower the threshold for use of nuclear weapons in a conventional conflict. Properly designed arms control could contribute to NATO security but appears unrealistic at this point. The more immediate challenge is how to deal with the fraying U.S.-Russia nuclear arms control regime.

The purpose of NATO's nuclear weapons

Since its founding in 1949, NATO has defined its primary responsibility as deterring and defending the territory and population of its member states against attack, in accordance with Article 5 of the Washington Treaty. Article 5 provides that:

> The Parties agree that an armed attack against one or more of them in Europe or North America shall be considered an attack against them all and consequently they agree that, if such an armed attack occurs, each of them,

> in exercise of the right of individual or collective self-defense recognized by Article 51 of the Charter of the United Nations, will assist the Party or Parties so attacked.[1]

NATO has since the 1950s regarded nuclear forces as central to its ability to deter attack. Successful deterrence requires persuading a potential adversary that the risks and costs of aggression far outweigh any possible benefits. Nuclear weapons pose huge risks and potential costs. For much of the Cold War, given significant Soviet and Warsaw Pact advantages in conventional armed forces, NATO overtly maintained the option to escalate to use of nuclear weapons in the event that its forces faced defeat during a conventional Soviet/Warsaw Pact.

Strategic nuclear forces provide the "supreme guarantee" of NATO security. The alliance's nuclear forces also include U.S. non-strategic nuclear weapons forward-deployed in Europe. They contribute to extended deterrence of potential adversaries, a problem that has posed and continues to pose particular challenges. U.S. strategic nuclear weapons provide a credible deterrent against attack on the United States; Russian (and, previously, Soviet) leaders would understand that such an attack would risk a nuclear conflict. The harder question has been to persuade Moscow (and assure allies) that an attack on one or more European NATO members would run the same hazard – that an American president would be prepared to risk Chicago for Berlin or The Hague.

Historically, U.S. and NATO officials have made a variety of attempts to address the extended deterrence challenge. These ranged from the aborted Multilateral Force in the early 1960s, which envisaged placing U.S. submarine-launched ballistic missiles on warships manned by multinational NATO crews, to the 1979 dual-track decision, which combined deployment of U.S. intermediate-range nuclear missiles in Europe to counter Soviet missiles with a U.S. effort to negotiate limits on such missiles, the end result being the 1987 Intermediate-range Nuclear Forces (INF) Treaty.

Beginning in the 1950s, the United States established programs of cooperation, under which U.S. nuclear weapons were maintained under the custody and control of U.S. personnel but in a conflict, with proper authorization, would be turned over to allies for use by their delivery systems. These "nuclear-sharing" arrangements are designed to foster NATO cohesion by spreading the responsibility for nuclear policy and posture as well as the risk. Forward-based U.S. nuclear weapons are seen as helping to make clear that a nuclear attack on one NATO member state could provoke a nuclear response by the alliance in accordance with Article 5. The presence of U.S. non-strategic nuclear weapons in Europe thus has as much, if not more, political value than military aspects.

Closely related to the extended deterrence challenge is that of assuring allies. By offering a physical manifestation of the U.S. commitment, U.S. non-strategic nuclear weapons in Europe are seen to bolster the assurance of allies. The Nuclear Planning Group, established in 1966 as the NATO body for planning and consultations on issues regarding nuclear policy and posture, also began as an effort to enhance ally assurances. The Nuclear Planning Group provided a venue in which European allies could feel that they "could significantly affect U.S. strategic nuclear policy without having physical access to the weapons themselves."[2] The Nuclear Planning Group supports the North Atlantic Council – which normally meets at the level of permanent representatives to NATO – the alliance's primary authority regarding NATO policy and posture.

During the Cold War, the contribution of U.S. forward-based nuclear weapons in Europe to extended deterrence and assurance seemed based in part on a belief that an American president would be more likely to authorize the use of non-strategic nuclear weapons against Soviet forces or the Soviet Union than to authorize the use of strategic nuclear weapons. One can question that but, fortunately, the proposition was never put to the test.

Assurance is not just a matter of specific weapons. It is also a matter of confidence among the alliance's leaders. If European leaders were absolutely convinced of the American president's readiness to defend them, including with strategic nuclear weapons, forward-based nuclear arms might not be necessary. Today, however, those weapons may have greater symbolic importance in view of U.S. President Donald Trump's erratic rhetoric of regard and behavior toward European allies, especially after the departure of U.S. Secretary of Defense Jim Mattis. President Trump's approach to NATO has caused consternation among allies, as did his decision to withdraw from the INF Treaty without prior consultation, and has led some to question the reliability of the U.S. nuclear security guarantee. He has provoked a (so far) low-key debate about how Europe might defend itself without the United States and a nascent discussion in Germany about what kind of nuclear cover that country might seek absent the extended U.S. nuclear deterrent.

The changing security environment in Europe

An era of promise

As the Cold War came to an end in the late 1980s, Europe's security environment radically transformed. West and East Germany reunified. The 1990 Charter of Paris for a New Europe reaffirmed the principles of the 1975 Helsinki Final Act and promised a new era of cooperation and friendly relations. The specter of Soviet military aggression melted away as 15 independent states emerged from the wreckage of the Soviet Union. Arms control played a major part in shaping this new security environment. The 1987 INF Treaty banned all U.S. and Soviet ground-launched cruise and ballistic missiles with ranges between 500 and 5,500 kilometers, not only in Europe but worldwide. The 1991 Strategic Arms Reduction Treaty (START, later referred to as START I) provided for significant reductions in U.S. and Soviet strategic nuclear forces.

In addition to negotiated reductions, the United States, Soviet Union and then Russia undertook unilateral steps in 1991 and 1992. President George H. W. Bush announced that the United States would eliminate all nuclear artillery shells and nuclear warheads for short-range missiles. That entailed the removal of 1,700 nuclear warheads from Europe. (NATO the following month agreed that the number of U.S. nuclear gravity bombs in Europe would be greatly reduced.) Moscow responded with unilateral nuclear reductions of its own.[3]

On the conventional side, the 1990 Treaty on Conventional Armed Forces in Europe (CFE Treaty) mandated drastic cuts in NATO and Warsaw Pact member-state holdings of tanks, armored personnel carriers, artillery, combat aircraft, and attack helicopters in the European region. As NATO prepared to enlarge, it engaged with Russia on developing a special relationship. That produced the 1997 Founding Act on Mutual Relations, Cooperation, and Security between NATO and the Russian Federation.

The Founding Act sought to assure Moscow that NATO enlargement would not bring a military threat closer to Russia. It stated,

> The member states of NATO reiterate that they have no intention, no plan and no reason to deploy nuclear weapons on the territory of new members, nor any need to change any aspect of NATO's nuclear posture or nuclear policy – and do not foresee any future need to do so.[4]

Regarding conventional forces, the document noted,

> NATO reiterates that in the current and foreseeable security environment, the alliance will carry out its collective defense and other missions by ensuring the necessary interoperability, integration, and capability for reinforcement rather than by additional permanent stationing of substantial combat forces.[5]

NATO communiqués reflected the positive trends in the security environment. The 1997 Madrid summit declaration focused on NATO enlargement and the parallel relationships with Russia and Ukraine. It stated that NATO leaders "continue to attach greatest importance to further the means of nonproliferation, arms control and disarmament" but was silent on the alliance's nuclear posture.[6] The organization marked its 50th anniversary with a summit in Washington in April 1999. The summit produced a new alliance strategic concept. The concept stated that the alliance had adopted "a significant relaxation of the readiness criteria for nuclear-roled forces" and that "NATO's nuclear forces no longer target any country." It added, however, that NATO "will maintain, at the minimum level consistent with the prevailing security environment, adequate sub-strategic [nonstrategic] forces based in Europe which will provide an essential link with strategic nuclear forces, reinforcing the transatlantic link."[7]

By 2008, some questioned the need for U.S. nuclear arms in Europe. The agreement of the governing coalition in Berlin called for the removal of U.S. nuclear weapons from Germany, while legislation was introduced in the Belgian parliament to ban the storage or possession of nuclear weapons on Belgian territory. Other allies, however, particularly those in Central Europe who had joined the alliance after 1997, argued that U.S. nuclear weapons should remain.[8] Some analysts argued for keeping U.S. nuclear weapons in Europe as their removal would greatly diminish the sharing of nuclear risk and responsibility within NATO by leaving the nuclear component of the alliance's deterrent based solely on strategic nuclear forces.

This debate was affected by NATO's nuclear legacy – the United States has had forward-based nuclear weapons in Europe since the 1950s. As one NATO member-state diplomat observed in 2010, "If there were no U.S. nuclear weapons in Europe today, NATO would not want to move them in, but moving them out is difficult due to the nuclear legacy."[9]

West-Russia relations fall to post-Cold War low

The positive developments in the European security environment did not last. Moscow became increasingly uneasy about NATO enlargement. Russian President Vladimir Putin delivered a strikingly strident speech at the February 2007 Munich Security Conference in which he took the United States to task for undermining European and global security, reflecting a view in the Russian government and security elite that the post-Cold War European order disadvantaged interests important to the Kremlin. Moscow in late 2007 "suspended" its observation of the CFE Treaty.

In 2008, Moscow objected vociferously to the prospect that NATO leaders, at their April Bucharest summit, might agree to give membership action plans to Ukraine and Georgia. In August 2008, Russia and Georgia fought a brief war, following which Russia recognized the break-away regions of Abkhazia and South Ossetia as independent states.

As the Russian economy and state budget strengthened, the Kremlin devoted significant resources to the modernization of Russian conventional, strategic, and non-strategic nuclear forces. The nuclear effort has included new ballistic missile submarines, new submarine-launched ballistic missiles (SLBMs), new intercontinental ballistic missiles (ICBMs) and reopening the Blackjack bomber production line. (See Chapter 3 for a more detailed discussion of Russian nuclear force modernization.)

President Barack Obama's attempted "reset" of the relationship with Russia seemed to stop the deterioration, at least for a few years. In an April 2009 speech in Prague, the president called for creation of conditions for a world free of nuclear weapons while reaffirming that, until a nuclear-free world could be achieved, the United States would maintain a safe, secure and effective nuclear deterrent.

The Obama administration's *Nuclear Posture Review*, released in April 2010, said the role of nuclear weapons in U.S. security policy had been reduced, adding, "The fundamental role of U.S. nuclear weapons, which will continue as long as nuclear weapons exist, is to deter nuclear attack on the United States, our allies, and partners." It went on to stress that the United States "would only consider the use of nuclear weapons in extreme circumstances to defend the vital interests of the United States or its allies and partners."[10]

The United States and Russia concluded the New START Treaty in 2010, which required that each side reduce to no more than 1,550 deployed strategic warheads and no more than 700 deployed strategic delivery vehicles. The Obama administration sought almost immediately to negotiate further reductions, including in non-strategic nuclear weapons. The Kremlin declined to engage on further cuts, however, citing the need to address questions such as missile defense, a long-standing issue of Russian concern. (The United States has not been prepared to discuss limits on missile defense since it withdrew from the 1972 Anti-Ballistic Missile Treaty in 2002.)

Despite the "reset," relations between Washington and Moscow cooled in 2012–13. Russia's use of military force to seize the Crimean peninsula from Ukraine in February 2014, followed in April by Russia's instigation of armed conflict in eastern Ukraine, led to a major downturn in relations between the United States, Europe, and Russia. Western countries began to apply various sanctions on Russia.

With growing concern about Moscow's intentions, NATO bolstered its conventional deterrent and defense capabilities in Central Europe, particularly in the Baltic region. NATO maintains, on a rotating basis, multinational battalion battle groups in Estonia, Latvia, Lithuania and Poland. In addition, the U.S. Army has deployed, also on a rotating basis, an armored brigade in Poland and elsewhere in Central Europe.

Concern grew further with Russia's aggressive use of cyber and social media to interfere in elections in the United States and Europe. Some analysts describe ongoing Russian "hybrid" actions – adversarial measures short of war – as aimed at sowing division within Western societies and weakening institutions such as the European Union and NATO.

Nuclear arms control in trouble

In July 2014, the U.S. government announced that Russia had violated the INF Treaty by developing and testing a ground-launched cruise missile to intermediate range. U.S. officials in early 2017 stated that Russia had deployed the prohibited missile. While Washington at first provided little detail about the offending missile, in late 2017 U.S. officials stated that the missile was the 9M729 cruise missile (NATO designator SSC-8).

Russian officials denied that they had violated the INF Treaty, claiming the 9M729 – a missile that they acknowledged only after the designator was disclosed by U.S. officials – had a range of less than 500 kilometers and thus was allowed. They charged that the United States had violated the treaty, in particular with the launcher system used for missile interceptors at the Aegis Ashore missile defense facility in Romania.

In October 2018, President Trump said the United States would withdraw from the INF Treaty. In early December, NATO foreign ministers stated, "Allies have concluded that Russia has developed and fielded a missile system, the 9M729, which violates the INF Treaty and poses significant risks to Euro-Atlantic security," though there was unhappiness that the Trump administration had presented them with a *fait accompli* on withdrawal. The ministers concluded by calling on "Russia to return urgently to full and verifiable compliance."[11]

As of January 2019, Russia had not returned to compliance. On February 1, the U.S. government announced that it was suspending its obligations under the INF Treaty and giving the required six-month notice of its intention to withdraw. The Russian government the next day suspended its INF Treaty obligations. On August 2, 2019, the United States formally withdrew from the INF Treaty.

Questions have also arisen about the fate of New START. While both countries met the New START limits when they took effect in February 2018, the treaty expires in February 2021. It can be extended by up to five years under its terms. Russia has expressed interest in discussing an extension, but as of mid-October 2019, the U.S. government had not decided whether to take up that question. The collapse of the INF Treaty and a failure to extend New START would mean that no limits would constrain U.S. and Russian nuclear forces, a dramatic change from most of the previous five decades that would decrease stability and security for the United States, Europe, and Russia.

Doctrinal questions

There is little doubt that Russia has shifted toward a more confrontational stance towards the West. That reflects a combination of factors: Russian domestic politics, Moscow's unhappiness with the existing European security order, a view in the Kremlin that Russia now has means to change the order, and an apparent confidence (or overconfidence) that, in a crisis or conflict with the West, Russia could threaten costs and risks that would force the West to back down.

Questions have arisen regarding Russia's nuclear doctrine and the possibility that Russia had lowered its threshold for use of nuclear weapons. The 2014 "Military Doctrine of the Russian Federation" seems benign on this question. It states that Russia reserves

> the right to use nuclear weapons in response to the use of nuclear and other types of weapons of mass destruction against it and/or its allies, as well as in the event of aggression against the Russian Federation with the use of conventional weapons when the very existence of the state is in jeopardy.[12]

The Russians, however, have offered no definition of what circumstances might put the "very existence" of the Russian state at stake. Some – including Pentagon and NATO analysts – believe Moscow might resort to nuclear weapons in circumstances other than an attack on Russia proper that faced the Russian military with the prospect of conventional defeat. The Russian military continues to maintain a large and diverse inventory of non-strategic nuclear weapons (NATO's nonstrategic inventory consists solely of U.S. B61 gravity bombs). Such considerations affected the Trump administration, whose 2018 *Nuclear Posture Review* says, "Russian strategy and doctrine emphasize the potential coercive and military uses of nuclear weapons. It mistakenly assesses that the threat of nuclear escalation or actual first use of nuclear weapons would serve to 'de-escalate' a conflict on terms favorable to Russia."[13]

Brad Roberts, a senior Department of Defense official during the Obama administration, expresses a similar concern. He postulates that Moscow has a theory of victory vis-à-vis the United States and NATO based on using conventional military forces to create a *fait accompli* on the ground in a regional conflict, combined with escalatory threats in which "non-strategic nuclear weapons apparently play a central role. . . . Actual employment would apparently be preemptive in nature and intended to de-escalate a conflict."[14]

Elbridge Colby, who served in the Department of Defense during the Trump administration, shares this view. He fears that, "In a contest with NATO, then, Russia might threaten to use or actually employ its nuclear forces in selective, tailored strikes to demonstrate Moscow's willingness to 'go nuclear' and thereby shock the alliance, break its political cohesion, and ultimately compel it to back down and terminate a conflict on terms favorable to Russia." Moscow would aim not to defeat NATO on the battlefield as much as to prevail in a contest of wills.[15]

Sometimes referred to as "escalate to de-escalate," the most common hypothetical example supposes that Russia launches a conventional attack in the Baltic states and enjoys initial success. However, when faced by a build-up of NATO conventional forces aimed at liberating the occupied territory, Russia escalates (or threatens to escalate) to the use of a small number of low-yield nuclear weapons to force NATO to back down. Some analysts worry that, given improving Russian air defense capabilities, U.S. and allied air forces might not be able to reliably deliver low-yield gravity nuclear bombs in response. With ICBM and SLBM warheads having yields of 100 kilotons or greater, they fear that an American president would be self-deterred from responding to a Russian attack with low-yield weapons, out of fear of triggering a large-scale strategic exchange.

To deal with these concerns, the Trump administration's *Nuclear Posture Review* describes a "tailored deterrence" strategy to ensure that Moscow understands that "any use of nuclear weapons, however limited, is unacceptable." To support this strategy, the review says the United States requires "a range of limited and graduated [nuclear] options, including a variety of delivery systems and explosive yields."[16] As discussed below, NATO policy also is focusing more on the nuclear component of the alliance's deterrent and defense posture.

Other analysts question whether the Department of Defense and NATO have correctly assessed Russian doctrine. Olga Oliker and Andrey Baklitsky contend that, while the Russians considered something like "escalate to de-escalate," it never made it into official doctrine. They note that neither the 2010 nor 2014 Russian military doctrine "in fact lowers the nuclear threshold . . . official statements, followed by a doctrine that did not deliver on them, suggest that proponents of a lowered threshold ultimately lost a bureaucratic fight."[17]

Kristin Ven Bruusgaard likewise argues that the logic in the 2018 *Nuclear Posture Review* "starts with a worst-case interpretation of Russian strategy and doctrine. There are indicators but no conclusive evidence to substantiate the claim that 'escalate to de-escalate' is Russian strategy." She writes that the new low-yield nuclear capabilities called for by the review "aim to solve a problem based on major and unsubstantiated assumptions about Russian doctrine. Such weapons will not meaningfully affect Russian calculations if the Kremlin fears the existence of their state is at stake."[18]

This author notes that, in a number of Track II discussions over the past several years in which he took part with nongovernmental Russian nuclear and arms control experts, every Russian save one denied that "escalate to de-escalate" was official Russian doctrine.

In October 2018, President Putin may have sought to dispel the notion that Moscow had lowered its threshold for employing nuclear weapons. He said,

> Our strategy of nuclear weapons use does not envisage a preemptive strike . . . Our concept is launch under attack. Only when we become convinced that there is an incoming attack on the territory of Russia, and that happens within seconds, only after that we would launch a retaliatory strike.[19]

However, these remarks seem geared to how Russia might respond to detection of a large missile attack against Russia, not how Moscow might consider use of nuclear weapons in a failing conventional conflict.

One other consideration regarding the debate over "escalate to de-escalate" and Russian doctrine should be borne in mind. Regardless of what a nuclear doctrine might say, in the end, a decision to escalate a conventional conflict by employing nuclear weapons, of whatever size, would be an intensely political decision. The leader making that decision would have to weigh the potentially ruinous consequences if the other side retaliates with nuclear weapons, which could set in motion an escalation spiral leading to catastrophe.

NATO nuclear posture

NATO's nuclear policy

The alliance's most recent strategic concept, issued in 2010, stated,

> Deterrence, based on an appropriate mix of nuclear and conventional capabilities, remains a core element of our overall strategy. The circumstances in which any use of nuclear weapons might have to be contemplated are extremely remote. As long as nuclear weapons exist, NATO will remain a nuclear alliance.[20]

The Deterrence and Defense Posture Review, issued at NATO's May 2012 summit, reflected similar themes. It stated, "Nuclear weapons are a core component of NATO's overall capabilities for deterrence and defense alongside conventional and missile defense forces." It added that allies "will ensure that all components of NATO's nuclear deterrent remain safe, secure and effective for as long as NATO remains a nuclear alliance."[21]

In the past five years, the alliance has begun paying more attention to its nuclear posture, particularly in view of concerns about Russia's large non-strategic nuclear arsenal, Russian nuclear doctrine, and the fact that Russia has significantly improved its conventional military forces and has regional superiority in certain regions, such as the Baltics. NATO's qualitative and quantitative advantages in conventional military forces have eroded since the 1990s and may continue to erode as Russia develops asymmetrical disruptive technologies.

Beginning with the September 2014 summit in Wales, NATO declarations began to address the Russian threat in more explicit terms, including the Russian nuclear threat. While the Wales summit declaration drew on the 2012 Deterrence and Defense Posture Review for its language on nuclear weapons and their contribution to the alliance's deterrence and defense capabilities, the communiqué from the July 2016 Warsaw summit and declaration from the July 2018 Brussels summit included more detailed language on nuclear weapons.

The Warsaw communiqué stated that three decision-making centers – Washington, London, and Paris – for the use of strategic nuclear forces "contribute to deterrence by complicating the calculations of potential adversaries." The communiqué noted the contribution of U.S. non-strategic nuclear arms in Europe (which had not been mentioned in the Wales declaration) and stated that the allies would ensure the "broadest possible participation of allies concerned in their agreed nuclear burden-sharing arrangements." It also cited the unique nature of nuclear weapons, warning that, "Any employment of nuclear weapons against NATO would fundamentally alter the nature of the conflict."[22]

The Brussels declaration two years later reiterated much of the Warsaw communiqué language. It contained new language on dual-capable aircraft, the delivery systems for U.S. non-strategic weapons: "National contributions of dual-capable aircraft to NATO's nuclear deterrence mission remain central to this effort [NATO's nuclear deterrence posture]." The declaration also stated that, "Allies concerned will continue to take steps to ensure sustained leadership focus and institutional excellence for the nuclear deterrence mission, coherence between conventional and nuclear components of NATO's deterrence and defense posture, and effective strategic communications."[23]

NATO strategic nuclear forces and modernization plans

The U.S. strategic nuclear triad consists of 14 Ohio-class ballistic missile submarines, each capable of carrying 20 Trident D5 SLBMs, 400 deployed Minuteman III ICBMs and some 60 B-2 and B-52H strategic bombers capable of carrying nuclear gravity bombs or air-launched cruise missiles (ALCMs). The United States is ramping up a program that will update all three legs of its strategic triad, including the production of the new Columbia-class ballistic missile submarine, a new ICBM, the B-21 strategic bomber and a new nuclear-armed ALCM, the Long-Range Standoff weapon or LRSO. The LRSO reportedly will be armed with a variable-yield warhead. Delivery of these new systems should begin in the second half of the 2020s. (See Chapter 2 for a more detailed discussion of current U.S. nuclear forces and modernization plans.)

There will almost certainly be debate in the United States, including in Congress, over the planned strategic modernization program and numbers. Part of the reason is cost. In October 2017, a Congressional Budget Office report projected the cost of U.S. nuclear weapons programs from 2017 to 2046 at $1.2 trillion (not adjusted for inflation and not counting the new weapons set out in the 2018 *Nuclear Posture Review*). However the discussion turns out, it is likely the United States will maintain a robust strategic triad.

British nuclear forces today consist solely of Trident D5 SLBMs on board four Vanguard-class ballistic missile submarines. (With the end of the Cold War, the British military eliminated its non-strategic nuclear weapons.) Each of Britain's ballistic missile submarines is armed with up to eight Trident D5 missiles, with each boat carrying up to 40 British-produced warheads.[24] The British use Trident D5s as part of a long-standing program of cooperation with the United States.

The British military currently maintains about 120 operationally available warheads, with a total arsenal of fewer than 215.[25] It is believed that some of the Trident warheads have been modified so that their yield is substantially reduced.[26] In 2015, the British government stated its intention to reduce to no more than 180 nuclear warheads in its stockpile by sometime in the mid-2020s.[27]

Britain bases its nuclear posture on the concept of minimum deterrence, the ability to inflict unacceptable damage on an adversary without having to match the size of the adversary's nuclear arsenal. During the Cold War, this was sometimes referred to as the "Moscow criterion" – the ability to destroy the Soviet political leadership and a few key industrial centers, including the Soviet capital.

Britain's SLBMs are committed to NATO's defense, but they remain operationally independent and could only be launched by order of the prime minister. This operational independence complicates an adversary's attack planning and enhances NATO deterrence, as the adversary would have to factor in the possible reaction of London as well as Washington. NATO's Nuclear Planning Group discusses plans for the employment of U.S. and British nuclear forces in a conflict, and the U.S. and British militaries reportedly coordinate on their nuclear contingency plans.

London has decided to replace its current ballistic missile submarines, which will start coming out of service in the late 2020s. In July 2016, the parliament voted 472 to 117 to proceed with the acquisition of new ballistic missile submarines. Labor Party leader Jeremy Corbyn voted against, but 140 members of the Labor Party joined the majority.[28] Britain plans to build four Dreadnought-class ballistic missile submarines to replace the Vanguard-class, with the first Dreadnought boat expected to go to sea in 2028. The new submarines will have 12 SLBM tubes each. The British Navy is working with the U.S. Navy to develop the Common Missile Compartment, which will be used on both the Dreadnought- and Columbia-class ballistic missile submarines to carry the extended-life version of the Trident D5 SLBM.[29]

French nuclear forces include SLBMs on board four Le Triomphant-class ballistic missile submarines. Each submarine can carry up to 16 M51 SLBMs. France in 2015 had 48 SLBMs that were operational (enough to equip three submarines), and each missile can carry up to six warheads.[30] France also maintains about 50 nuclear-capable Rafale F3 fighter-bombers that can carry nuclear-armed ASMP-A ALCMs. Many analysts would consider the fighter-bombers and their nuclear armament to be non-strategic nuclear weapons, but the French government views them as strategic. France is estimated to have a total inventory of about 300 nuclear weapons, currently making up the third largest nuclear arsenal after those of Russia and the United States.

French doctrine states that the rationale for French strategic nuclear forces is to guarantee the sovereignty of France. Then-President Hollande said in 2015 that nuclear forces would be used to target "adversary centers of power" but not population centers. While NATO regards French nuclear forces as contributing to alliance security, Paris has not committed its nuclear forces to NATO. French representatives do not participate in the Nuclear Planning Group, though French diplomats coordinate with their U.S. and British counterparts on nuclear language for summit communiqués. As with the case of Britain, however, the existence of an independent French decision point on the possible use of nuclear weapons in a conflict involving NATO is seen as enhancing deterrence.

French nuclear forces have been recently modernized, with the M45 SLBM replaced by the longer-range M51 missile and the Mirage 2000Ns phased out and replaced by Rafale F3s. France is developing an upgraded version of its M51 SLBM, the M51.2, and is beginning to plan for a next-generation ballistic missile submarine, to enter service in the mid-2030s. It is also considering the weapon that will replace the ASMP-A for its Rafale F3 aircraft, possibly an air-launched hypersonic missile.[31]

NATO non-strategic nuclear forces and modernization plans

During the Cold War, the United States for purposes of extended deterrence and assurance maintained a large number of non-strategic nuclear weapons of various kinds in Europe. These included gravity bombs, artillery shells, landmines, warheads for short- and intermediate-range missiles and surface-to-air missiles, and depth charges. The number of U.S. nuclear warheads in European NATO countries peaked at around 7,300 in the early 1970s. Some 2,800 of those were designated for use by allied delivery systems under programs of cooperation.[32]

Currently, the United States has only one non-strategic nuclear weapon in its arsenal: the B61 gravity bomb. The U.S. Air Force reportedly has 300 of these weapons for delivery by dual-capable aircraft. Some 150 are forward-deployed in Europe at six locations. Eighty of these weapons are designated for possible use by NATO allies under programs of cooperation. B61 gravity bombs are stored at Aviano Air Base in Italy for use by U.S. F-15s and F-16s. B61 bombs are also deployed at Kleine Brogel Air Base in Belgium, Volkel Air Base in the Netherlands, Buechel Air Base in Germany, Ghedi Air Base in Italy and Incirlik Air Base in Turkey. In a conflict, these weapons could be turned over to allies and delivered by Belgian and Dutch F-16s and by German and Italian PA-200 Tornados; the status of the Turkish delivery capability is unclear.[33]

The United States is modernizing and extending the life of the B61 bomb. The modernization plan is taking nonstrategic variants of the bomb and the strategic variant B61-7 and will result in the B61-12, which will be used to arm both strategic bombers (B-2 and B-21) and dual-capable aircraft (F-15, F-16 and F-35 Joint Strike Fighter). The B61-12, which replaces four different types of the B61 bomb that could produce different weapons effects, is estimated to have a variable yield between 0.3 and 50 kilotons. It will be equipped with a tail-kit guidance assembly that will make the bomb far more accurate than its predecessors, though the tail-kit may not be usable with the F-16 and Tornado. Serial production is slated to begin in early 2020 and will result in 400–500 bombs when it concludes in the mid-2020s.[34]

The program should extend the life of the B61 bomb by 20–30 years. The program's estimated cost in 2016 was $7.6 billion (a considerable increase on original estimates, though some projected the cost to rise to as high as $10 billion). The new tail-kit for the bombs will add an additional estimated $1.3 billion to the cost.[35]

Delivery systems for non-strategic nuclear weapons also are being modernized. The U.S. Air Force plans that Block 4 of the F-35A will be nuclear-certified to carry the B61-12. Those F-35As are expected to begin joining the U.S. Air Force in the early 2020s.

Some in 2009–2010 predicted that non-U.S. NATO air forces would lose their nuclear delivery capability as dual-capable F-16s and Tornados were retired from service, at about the same time as suggestions that U.S. forward-based nuclear weapons could be returned to the United States. That has changed, in part reflecting concern about Russia's nuclear and overall military posture. Italy, the Netherlands, Turkey and Belgium, the last just in October 2018, have decided to purchase F-35As to replace their aging dual-capable aircraft. Thus, four of the five NATO countries that currently host U.S. nuclear weapons are on track to acquire aircraft capable of delivering the B61-12.

The fifth is weighing its options. The German Ministry of Defense believes that its Tornados should be phased out in the mid-2020s. Among the replacement options is a dual-capable Typhoon Eurofighter. The Typhoon is not certified to carry nuclear weapons, and German officials reportedly approached U.S. officials in spring 2018 to ask what would be necessary to certify the Typhoon.[36] A possibility for the longer term is a next generation fighter aircraft that Germany seeks to develop in cooperation with France.

Other U.S. weapons?

Some discussion has begun in Europe over whether, once it has suspended its obligations under the INF Treaty, the United States might propose to deploy a new intermediate-range missile, perhaps a ground-launched cruise missile, in

Europe. The speculation focuses on a conventionally-armed missile, but even that likely would prove controversial within NATO. Most U.S. discussion of a new intermediate-range missile has focused on China and the Pacific region rather than Europe. Department of Defense officials in March 2019 said that they plan to test two ground-launched intermediate-range missiles later in the year: a 1,000-kilometer range ground-launched cruise missile based on the Tomahawk sea-launched cruise missile, and a ground-launched ballistic missile with a range of 3,000–4,000 kilometers.[37]

Questions have arisen in the United States regarding new nuclear weapons capabilities. The Trump administration's *Nuclear Posture Review* notes that Russia is modernizing its stockpile of some 2,000 non-strategic nuclear weapons and calls for "supplements" to the Obama administration's planned nuclear modernization program in order "to enhance the flexibility and responsiveness of U.S. nuclear forces."[38]

Specifically, the Pentagon plans to supplement current non-strategic nuclear weapons systems – B61 gravity bombs delivered by dual-capable aircraft – with two new capabilities. These will be a low-yield warhead for a small number of Trident D5 SLBMs and a modern sea-launched cruise missile with a low- or variable-yield nuclear capability. The *Nuclear Posture Review* noted that these would add diversity to U.S. nuclear platforms, enhancing the ability of the U.S. nuclear deterrent to deal both with current challenges and future "breakout" scenarios.

The attraction for the Trump administration of a low-yield variant of the W76 appears to be two-fold. First, it will require a modification of existing W76 warheads that can be accomplished relatively quickly and at comparatively little cost. Indeed, the first modified warhead was reportedly delivered in early 2019. Second, as the Russians (and others) have developed increasingly sophisticated air defenses, concern has grown about the ability of aircraft and cruise missiles to reliably penetrate those air defenses. A low-yield warhead delivered by a Trident D5 would circumvent air defenses and likely defeat missile defense systems designed to deal with shorter-range ballistic missiles. (This would give the W76 a new purpose for a warhead previously seen as strategic.)

Developing and deploying a nuclear-armed sea-launched cruise missile will require more time and prove more expensive. In 1991, all nuclear-armed Tomahawk SLCMs were removed from U.S. Navy warships. The W80-0 nuclear warheads were placed in storage, and the Tomahawk missiles were retired or converted to carry conventional warheads. While the Obama administration retired the W80-0 warhead, some version of the W80 will apparently be converted for the new SLCM. A nuclear-armed SLCM may contribute more to extended deterrence and assurance in the Pacific region, where – unlike Europe – no U.S. non-strategic nuclear weapons are based.

Senior Pentagon officials have held out the possibility that the U.S. government would reconsider a nuclear-armed sea-launched cruise missile if Russia returns to compliance with the INF Treaty. It is unclear, however, whether Moscow would be prepared to meet the more extensive requirements laid out in the Trump administration's *Nuclear Posture Review*, which states, "If Russia returns to compliance with its arms control obligations, reduces its non-strategic nuclear arsenal, and corrects other [unspecified] destabilizing behaviors, the United States may reconsider the pursuit of a SLCM."[39]

The new weapons will face scrutiny in Congress. Congress voted to provide funds for the low-yield W76 warhead in the fiscal year 2019 budget but did so over objections from Democrats in the House of Representatives. The challenges to the warhead will likely intensify with the Democrats now holding the majority in the House.

The Pentagon will likely also face a debate in Congress when it requests funding for a nuclear-armed sea-launched cruise missile. Opponents will question what added value a nuclear-armed SLCM would bring when the United States in the early 2020s will be deploying stealthy F-35s capable of delivering modernized B61-12 nuclear bombs (with yields as low as 0.3 kilotons), to say nothing of cruise missiles that could be launched by U.S. strategic bombers.

For the near-term, this debate over a low-yield Trident D5 warhead and nuclear-armed SLCM likely will be a debate within the United States. Since the systems are sea-based, there was no requirement for NATO approval, and no evidence suggests that U.S. officials consulted in any detail with allies on these systems prior to completion of the 2018 *Nuclear Posture Review*. However, if these weapons are built and added to the U.S. arsenal, there will be discussions – and possibly some debate – within the Nuclear Planning Group on how they fit into NATO's deterrence and defense posture.

No nuclear weapon has been used in conflict for more than 70 years, giving rise to what is sometimes referred to as the "nuclear taboo." It is in the interest of the United States and NATO to maintain this taboo and to make clear that, if the nuclear threshold is ever crossed, the game will dramatically change, and the consequences will become unpredictable and potentially catastrophic. However, if Russia, the United States, and NATO increase reliance on low-yield or variable nuclear weapons in their force postures and doctrines, they risk – perhaps inadvertently – lowering the nuclear threshold and/or risk blurring that threshold by somehow suggesting that use of "small" nuclear weapons is different and permissible. That would not be in the West's interest.

The most likely scenario that would produce a nuclear exchange between NATO and Russia is the escalation of a conventional conflict. It would be in NATO's interest to maintain conventional forces that could deter any conventional conflict to begin with. That will require sustained defense investments by alliance members, but by reducing the risk of conventional conflict, they would also reduce the risk of a nuclear exchange.

A need for dialogue

In the aftermath of Russia's military seizure of Crimea and use of force to provoke and sustain a simmering conflict in eastern Ukraine, Washington and NATO suspended many channels of communication with Moscow to send a message of no business as usual. In the current circumstances, reestablishing military-to-military and other channels to address nuclear and other security issues makes eminent sense.

Following the signing of the New START Treaty, many hoped that Washington and Moscow would move rapidly to a second agreement that would further reduce strategic nuclear arms and bring non-strategic nuclear weapons under control. Those hopes did not materialize. It would be in the interest of NATO and European security to get to a point where negotiations address reductions in non-strategic nuclear arms, particularly the excessive number of Russian non-strategic weapons.

Given the current political climate, however, there appear few prospects for serious arms control in the near term. Quite the opposite, as it appears that, on its current course, the INF Treaty will become a dead issue in 2019 and beyond. An arms control dialogue should aim first at preserving the existing nuclear arms control regime, that is, extending New START and, if at all possible, promoting compliance and rescusitation of the INF Treaty.

Whether or not some kind of arms control dialogue can be restarted, NATO and the United States on a bilateral basis should continue to try to engage Russia in detailed discussions of issues such as minimizing the risk of accident or miscalculation when U.S./NATO and Russia's military forces operate in close proximity and on correctly understanding the sides' nuclear doctrines. Moscow does not appear eager for such a discussion. It nevertheless should be in the interest of both sides to avoid miscalculation or incorrect assessments of the other's doctrine.

These would be logical topics for a NATO-Russia channel or a bilateral U.S.-Russia channel, such as the strategic stability talks. Unfortunately, however, the strategic stability talks held only one brief session in 2017 and did not meet in 2018 or early 2019. Moscow might not be prepared to engage seriously, but the United States and NATO lose nothing by trying. The absence of dialogue raises the odds of miscalculation and that NATO and Russia will fall back on worst-case assumptions as they make decisions regarding their nuclear postures and modernization programs. Invariably, those decisions will be more costly and could result in a less stable and less secure European security environment.

Notes

1 NATO, "The North Atlantic Treaty," Washington, DC, April 4, 1949, www.nato.int/cps/ie/natohq/official_texts_17120.htm.
2 David N. Schwartz, *NATO's Nuclear Dilemmas* (Washington, DC: Brookings, 1983), 185.
3 Susan J. Koch, "The Presidential Nuclear Initiatives of 1991–1992," Case Study #5, Center for the Study of Weapons of Mass Destruction, National Defense University, September 2012, https://wmdcenter.ndu.edu/Portals/97/Documents/Publications/Case%20Studies/cswmd_cs5.pdf.
4 North Atlantic Treaty Organisation (NATO), Founding Act on Mutual Relations, Cooperation and Security between NATO and the Russian Federation signed in Paris, France, 1997, https://www.nato.int/cps/en/natohq/official_texts_25470.htm?selectedLocale=en.

5 NATO, "Founding Action Mutual Relations, Cooperation and Security Between NATO and the Russian Federation Signed in Paris, France," May 27, 1997, www.nato.int/cps/su/natohq/official_texts_25468.htm.
6 NATO, "Madrid Declaration on Euro-Atlantic Security and Cooperation Issued by the Heads of State and Government," July 8, 1997, www.nato.int/docu/pr/1997/p97-081e.htm.
7 NATO, "The Alliance's Strategic Concept, Approved by the Heads of State and Government participating in the meeting of the North Atlantic Council in Washington, DC," April 24, 1999, www.nato.int/cps/ie/natohq/official_texts_27433.htm.
8 Steven Pifer, "NATO, Nuclear Weapons and Arms Control," Brookings Arms Control Series Paper 7, July 2011, 11–12.
9 Author's conversation with Norwegian diplomat at NATO headquarters, November 2010.
10 Office of the Secretary of Defense, "Nuclear Posture Review, April 2010," 15–16, https://dod.defense.gov/Portals/1/features/defenseReviews/NPR/2010_Nuclear_Posture_Review_Report.pdf.
11 NATO, "Statement on the Intermediate-Range Nuclear Forces (INF) Treaty, Issued by the NATO Foreign Ministers," Brussels, December 4, 2018, www.nato.int/cps/en/natohq/official_texts_161122.htm.
12 Theatrum-Belli, "The Military Doctrine of the Russian Federation, Translated from the Russian," June 29, 2015, https://theatrum-belli.com/the-military-doctrine-of-the-russian-federation/.
13 U.S. Defense Department, Office of the Secretary of Defense, "*Nuclear Posture Review* 2018," February 2018, 8, https://fas.org/wp-content/uploads/media/2018-Nuclear-Posture-Review-Version-2.pdf.
14 Brad Roberts, *The Case for U.S. Nuclear Weapons in the 21sst Century* (Stanford, CA: Stanford University Press, 2016), 128–138.
15 Elbridge Colby, "Countering Russian Nuclear Strategy in Central Europe," *Center for a New American Security*, November 11, 2015, www.cnas.org/publications/commentary/countering-russian-nuclear-strategy-in-central-europe.
16 U.S. Defense Department, Office of the Secretary of Defense, *Nuclear Posture Review* 2018, 30–31.
17 Olga Oliker and Andrey Baklitsky, "The *Nuclear Posture Review* and Russian 'De-Escalation:' A Dangerous Solution to a Nonexistent Problem," *War on the Rocks*, February 20, 2018, https://warontherocks.com/2018/02/nuclear-posture-review-russian-de-escalation-dangerous-solution-nonexistent-problem/.
18 Kristin Ven Bruusgaard, "Here's Why U.S. Tactical Nuclear Weapons Are a Bad Idea: They Likely Won't Change Moscow's Calculations During a Crisis," *The National Interest*, December 10, 2018, https://nationalinterest.org/feature/heres-why-us-tactical-nukes-are-bad-idea-38372.
19 Mike Eckel, "Putin Says Russia Has Ruled Out Launching Preemptive Nuclear Strikes," *Radio Free Europe/Radio Liberty*, October 19, 2018, www.rferl.org/a/putin-says-russia-has-ruled-out-launching-preemptive-nuclear-strikes-only-use-defensively/29552055.html.
20 NATO, "Active Engagement, Modern Defence: Strategic Concept for the Defense and Security of the Members of the North Atlantic Treaty Organization, Adopted by Heads of State and Government at the NATO Summit in Lisbon," November 19–20, 2010, www.nato.int/nato_static_fl2014/assets/pdf/pdf_publications/20120214_strategic-concept-2010-eng.pdf.
21 NATO, "Deterrence and Defense Posture Review," issued on May 20, 2012, www.nato.int/cps/en/natohq/official_texts_87597.htm.
22 NATO, "Warsaw Summit Communiqué Issued by the Heads of State and Government Participating in the Meeting of the North Atlantic Council in Warsaw 8–9 July 2016," July 9, 2016, www.nato.int/cps/en/natohq/official_texts_133169.htm.

23 NATO, "Brussels Summit Declaration Issued by the Heads of State and Government Participating in the Meeting of the North Atlantic Council in Brussels 11–12 July 2018," July 11, 2018, www.nato.int/cps/en/natohq/official_texts_156624.htm.
24 "National Security Strategic and Strategic Defence and Security Review 2015," UK Parliament, 34, www.gov.uk/government/uploads/system/uploads/attachment_data/file/478933/52309_Cm_9161_NSS_SD_Review_web_only.pdf.
25 Hans Kristensen and Robert Norris, "Global Nuclear Weapons Inventories, 1945–2010," *Bulletin of the Atomic Scientists*, July 2010, 81, http://bos.sage-pub.com/content/66/4/77.full.pdf+html.
26 Michael Quinlan, *Thinking About Nuclear Weapons: Principles, Problems, Prospects* (Oxford: Oxford University Press, 2009), 127.
27 "National Security Strategic and Strategic Defence and Security Review 2015," 34.
28 "MPs Vote to Renew Trident Weapons System," *BBC News*, July 19, 2016, www.bbc.com/news/uk-politics-36830923.
29 George Allison, "A Guide to the Dreadnought-Class Ballistic Missile Submarine," *UK Defence Journal* (October 24, 2017), https://ukdefencejournal.org.uk/guide-dreadnought-class-ballistic-missile-submarine/.
30 "France," *Nuclear Threat Initiative*, May 2016, www.nti.org/learn/countries/france/nuclear/.
31 Steven Pifer and James Tyson, "Third-Country Nuclear Forces and Possible Measures for Multilateral Arms Control," Brookings Institution, August 2016, 21–22, www.brookings.edu/wp-content/uploads/2016/08/acnpi_20160824_multilateral_arms_control_01.pdf.
32 Robert Standish Norris, "United States Nuclear Weapons Deployments Abroad, 1950–1977," *Carnegie Endowment for International Peace*, November 30, 1999.
33 Hans M. Kristensen and Robert S. Norris, "United States Nuclear Forces, 2018," *Bulletin of the Atomic Scientists* 74, no. 2 (2018), 120–131.
34 Joseph Trevithick, "Getting to Know America's Long Serving B61 Family of Nuclear Bombs," *The War Zone*, March 15, 2018, www.thedrive.com/the-war-zone/19263/get-to-know-americas-long-serving-b61-family-of-nuclear-bombs.
35 Kingston Reif, "U.S. Nuclear Modernization Programs," *Arms Control Association*, August 2018, www.armscontrol.org/factsheets/USNuclearModernization.
36 Joseph Trevithick, "The German Air Force Wants to Know If Its Eurofighters Can Carry U.S. Nuclear Bombs," *The Drive*, June 21, 2018, www.thedrive.com/the-war-zone/21679/the-german-air-force-wants-to-know-if-its-eurofighters-can-carry-u-s-nuclear-bombs.
37 Shervin Taheran, "U.S. to Test INF Treaty-Range Missiles," *Arms Control Association*, April 1, 2019, www.armscontrol.org/act/2019-04/news/us-test-inf-treaty-range-missiles.
38 U.S. Defense Department, Office of the Secretary of Defense, *Nuclear Posture Review* 2018, 52.
39 Ibid., 55.

8 Through a crystal ball, dimly

Nuclear modernization's anticipated effects on International Relations theory

Balazs Martonffy and Eleni Ekmektsioglou

Nukes, crystal balls, and futures of theories

The ivory tower of academe may be intricate in nature, byzantine in design, and daedal in style, but it does not come equipped with a working crystal ball. Within the world of International Relations (IR) theory, predictions are hard, especially about the future. Rarely do we find theories that go beyond their descriptive, evaluative, and explanatory role to not only demonstrate a causal relationship, but attempt to make probabilistic or deterministic claims about the future. Even rarer is it to find scholarly work on how theory itself will evolve, unless some evidentiary fiat is so *interesting* that future-finding transcends the daydream and becomes a justifiable and warranted analytical exercise. The advent of the nuclear age undeniably was one such fiat, and nuclear modernization is another such directive, one that asks us to gaze into the future of how theory might respond to incremental changes in the empirical world. While we will not be able to predict the winding path theory will take, we will probe, explore, and plot out a road-map where theory development will likely find solid footing.

Kenneth Waltz, the father of modern International Relations theory, explicitly stated that, theory-making lies in looking for simplicities while taking a step back from reality.[1] For years, International Relations theory has been aspiring to parsimony at the expense of empirically driven research and fine-grained theories with limited generalizability. The evolution of the discipline notwithstanding, compelling theoretical arguments about international phenomena are inextricably linked to empirical observations. Put differently, International Relations theory remains an empirically driven scientific proposition while the iterative nature and dialectical relationship between deductive and inductive theorizing is inescapable.

More specifically, and with regard to nuclear weapons and state relations, the current nuclear modernization efforts have the potential to bring about changes in the currently somewhat stale empirical realm of nuclear studies, and may even jolt novel, more general discussions about International Relations theoretical schools of thought and paradigms. This chapter is not concerned with an extensive or intensive analysis of the technological or policy implications of the ongoing and proposed future U.S. nuclear modernization plans. Instead, it is interested in providing an overview of what sort of effects nuclear modernization would bring about in International Relations theory.

Nuclear modernization then, as previous chapters have contextualized, is taken as a potential micro-driver of theoretical change and examined from that perspective. While the counter positive of this type of analytical reasoning, that of the outcomes of the double hermeneutical effect, i.e. when theory informs policy changes by affecting the mindset of the policy-maker, are amply determined, it is not unreasonable to expect new empirical evidence to generate theoretical changes as well. We examine this possibility.

In the first section of this chapter, we offer an overview of the main Cold War theories and the way they conceptualized the impact of nuclear weapons on state interactions. The second section explores ways the current U.S. modernization might impact traditional macro-level theories of International Relations and we, then, proceed by considering new evidence regarding middle-range theories and their potential re-consideration in the post-Cold War environment. The last section looks at some critical foreign policy questions such as the U.S.-China relationship and the impact of the U.S. modernization on theories of nonproliferation.

Cold War theories: existential deterrence against nuclear war fighting and damage limitation

When Barnard Brodie called the nuclear weapon 'the absolute weapon', he argued that war would never be the same again after the explosion of the atomic bomb.[2] The massive destructiveness of the new weapon completely transformed the way states had been fighting until then. Some scholars went as far as to say that war was obsolescence – an argument that was further reinforced by the hydrogen bomb, given that nuclear powers appeared to be able to promise total destruction to their adversary through the use of their nuclear arsenal.[3] In other words, the first nuclear strategists of the Cold War era regarded nuclear weapons as capable of deterring both nuclear and conventional threats; just the existence of nuclear weapons in a state's arsenal was the absolute deterrent of both conventional and nuclear conflicts. The main logic at the heart of the nuclear strategy argument was related to the inherent escalatory dynamics of every conflict, which after the invention of nuclear weapons, resulted in being prohibitively costly for any given benefit. That said, the first generation of nuclear theorists identified a positive effect of the bomb on state behavior. For Thomas C. Schelling, Bernard Brodie, or Albert Wohlstetter, the bomb pushed states towards self-restraint and bargaining rather than the use of force.

The above-mentioned theoretical insights paved the way for a new generation of scholars that argued for the numerical irrelevance of nuclear weapons as far as their deterrent power is concerned. The stabilizing effect of nuclear weapons is regarded as robust and non-dependent on numerical or technical superiority. The existential deterrence thesis argues that the destructive capability of even just one nuclear warhead is so great, that conventional conflict needed to be avoided even if the prospect of nuclear retaliation appears to be small.[4]

Middle range theories of deterrence influenced macro-level theories of International Relations. For scholars such as Kenneth Waltz, one nuclear bomb carried enough deterrent power to prevent adversaries from using their conventional or nuclear arsenals against the state.[5] Scholarship regarded all nuclear weapons states as having the same deterrent power emanating from a single nuclear weapon or a big arsenal; numbers were irrelevant as much as secured second-strike capabilities were.[6] For theorists like Waltz, the nuclear weapon was the absolute force equalizer and the cost imposed by its use exceeded any benefit gained from a conflict. Nuclear weapons have a stabilizing effect on state relations. On the other hand, Glenn Snyder's stability/instability paradox concept provided the theoretical foundation for the pessimists of the debate to argue that nuclearization did not prevent states from initiating conventional conflicts but, on the contrary, created the space for weaker conventionally states to engage in limited warfare under the nuclear umbrella.[7] Escalatory dynamics during a conventional conflict pushed thinkers towards concepts of warfighting even between two nuclear powers. The need to capitalize on nuclear weapons and their military usability as well as political relevance, was reinforced by the conventional inferiority of the U.S. vis-à-vis the Soviet Union and the U.S. need to reassure its allies on its commitment to defend them.[8] Existential deterrence was challenged by theories of warfighting doctrines and counterforce postures against the enemy's command and control centers along with critical nodes of its nuclear arsenal.[9]

In a nutshell, during the Cold War, nuclear strategy theorists and practitioners vacillated between seeing the world through either a lens that put the emphasis on Mutually Assured Destruction (MAD)/mutual vulnerability through deterrence by punishment, or a lens that focused on damage limitation and denial of retaliatory strikes through counterforce postures. In the immediate post-Cold War environment, however, great power competition faded. The influence of nuclear weapons on world politics seemed to also fade out, and the general trend of continued drawdown between the nuclear superpowers since the latter part of the Cold War has been pursued by both Democratic and Republican administrations and Moscow as well. The current U.S. administration's nuclear modernization plans, however, which include modernizing strategic delivery systems, production complexes, and general force improvement as well as low yield nuclear warheads,[10] create questions about the goals the U.S. is trying to achieve through its current modernization and their impact on deterrence and future nuclear politics. In fact, some scholars have argued that innovation in nuclear weapons has transformed nuclear strategy rendering counterforce options much more feasible than in the past.[11] In essence, what counterforce advocates have been asking for is a U.S. policy of seeking nuclear supremacy through a series of investments in accurate and prompt delivery systems, new reconnaissance technologies, and smart warheads of flexible yield.[12] Counterforce advocates conclude that new technologies make the use of nuclear weapons a real policy alternative given their high accuracy, which translates in low numbers of fatalities and collateral damage.

At the same time, the nuclear modernization plan of the Russian nuclear arsenal puts the emphasis on new ground-based delivery systems, nuclear-powered cruise missiles, and hypersonic weapons whose objective is to penetrate missile defenses and deliver nuclear strikes promptly.[13] The high accuracy of both cruise and hypersonic missiles pushes one to assume that the Russian nuclear strategy is becoming gradually more counterforce-focused than countervalue. Even though the articulated objective of these new weapons is to restore strategic stability with the U.S. after the introduction of missile defenses and deter a counterforce strike from the U.S., the weapons' offensive potential and uncertainty felt within U.S. military and civilian circles has led many to argue that a new arms race is inevitable.[14]

While the extent to which counterforce strategies are feasible today is still a topic to be discussed further, one could still wonder how nuclear politics can be theorized in the post-Cold War era where the gap of military capabilities between the U.S. and other nuclear powers keeps getting wider, which motivates states like China and Russia to find ways to catch up in order to counter U.S. advantages.[15] In the following sections, we explore possible ways nuclear modernization in the U.S. could influence macro-level along with middle range theories and vice versa. We regard the relationship between theories and policy-making as dialectical and we explore how they can interact within the context of the U.S. nuclear modernization plans, comparing the Cold War to the post-Cold War international environment.

Micro-drivers of macro-theories: realism vindicated?

The origins of modern International Relations theory began with the devastation brought about by World War II, when classical realist scholars such as Hans Morgenthau were delineating international politics as a separate field from economics, law, and other social sciences.[16] While empirically founded initially, most of International Relations theory by now is well formed and remains spatially and temporally substantive, withstanding over seventy years of continuous academic scrutiny. Given this stability in the field, empirical micro-drivers are unlikely to have major, game-changing effects on International Relations theory as a whole. Even so, nuclear modernization is a unique set of policy choices by the United States that may bring about the empirical impetus for a possibility for theory to evolve, adapt, be refined, and be further expanded on in a myriad of ways.

Our overarching assertion is that the fundamentals of International Relations theory will not be challenged by nuclear modernization, as the field remains resistant to single evidentiary challenges. But the possibility that the current U.S. nuclear modernization proposals could fundamentally alter International Relations theory does exist, with the following caveat: if the ongoing nuclear modernization is not merely a single event or set of events, but a cascading series of nuclear build-ups, resulting in a major shift in the current and past proliferation trends. If this current set of doctrine and nuclear build-up results in major increases in both horizontal and vertical proliferation, theory will change in a substantial way. This avenue of thought and theoretical argumentation, while certainly plausible, remains to be assessed at a future time.

Further, the ontological and epistemological foundations of International Relations theory remain fundamentally unaffected. Nuclear modernization does not bring about a teleological or philosophical shift in the nature of social reality from an objective viewpoint (as much as objectivity can be an attribute assigned to social reality), nor is it likely to be perceived as a fundament shift in epistemology. The divisions within our discipline, short-handedly detailed as positivist and interpretivist research, are unlikely to come closer as their fundamental underpinnings remain far apart. Interpretivist research will likely focus on identifying the underlying hegemonic discourse within nuclear modernization, while positivist research, writ large, will attempt to incorporate this new evidence into existing theory, focusing on novel theoretical causal chains, linkages, and mechanisms.

Other far-reaching effects on ontological and epistemological considerations in International Relations theory, except the ones outlined above, are unlikely, but the concept of security is ever changing. Nuclear modernization, as an important policy decision, may challenge the concept of security itself. Interesting debates about the intersections of national security, state security, and human security are bound to surface. In the same vein as Krause and William's seminal 1996 piece,[17] which attempted to link the debates between neorealist and critical approaches to security, a novel re-broadening of the security studies agenda may see sunlight. Yet even if such an endeavor were to take place, it is unlikely that the entire revision of the concept of security itself, as Baldwin did in 1997[18] in his work that expanded the conceptualization of the term 'security', will happen within our discipline in consequence.

Leaving conceptual debates behind, the next query concerns changes that may be brought about in what the disciple identifies as schools of thought, or paradigms within International Relations. While divisions exist within each broad "house" of thought, the three commonly accepted ones, as detailed by Stephen Walt, are realism, liberalism, and constructivism.[19] While these paradigms are also very resistant to change, there is a natural ebb and flow in their pattern of recognition in the "marketplace of ideas" that the pages of the top journals in our field reflect.[20] It is undeniable that certain events bring one of the three more limelight than others. For example, Brexit debates bring to the forefront questions of European Union integration and cohesion, which earlier we thought settled, and liberal ideas are being tested at the writing of this article. The simmering debate between the role, place, and conceptual validity of power, reciprocity, and norms and ideas continues to be debated as new evidence surfaces. Nuclear modernization will assuredly challenge and test these paradigms' applicability today.

Reinforcing constructivist approaches, the non-use of nuclear weapons after World War II provided a substantial empirical starting point for constructivist research. Perhaps the most notable of these works in the nuclear studies field is Nina Tannenwald's nuclear taboo,[21] and follow-on theoretical work on nuclear modernization will for sure build on this. Constructivists will examine the challenges to the enduring role of established institutions and explore the possibility of norms weakening around the nuclear proliferation taboo.

The fall of the Soviet Union in 1991 left the United States as a global hegemon and a unipole from an International Relations theoretical perspective in world politics. Established realist theory of the time asserted that a coalition of other states, fearing for their security, were sure to balance against the United States. Empirics clearly did not match theoretical expectations, and the behavior of the United States continued to puzzle scholars. The concept of a benign hegemon, the assumption of unipolarity, and balancing against threat or threat perception were theorized. A great power that begins to heavily modernize may reignite this debate. In addition, the debates that are currently simmering about the role of threat versus its perception, the effect of misperceptions, inadvertent and conscious biases, may all be revisited under different conditions, especially as nuclear modernization takes off. Similarly, the European Union's success as an integration model of supra-national organization provided generalizable theoretical works on inter-state cooperation. Nuclear modernization is most likely not of this caliber in providing fuel for liberal scholarly work, but the role that successful extended nonproliferation cooperation systems or regimes, such as the Proliferation Security Initiative,[22] the Wassenaar Arrangement,[23] or the Zangger Committee,[24] may be further examined from a theoretical standpoint incorporating the novel evidence. Further, the Nuclear Nonproliferation Treaty, a long-standing staple of a successful arms control agreement, will come under heavy scrutiny as nuclear modernization begins in earnest. The possible lagged effects of forced cooperation and the time-horizon questions of inter-state treaties will both be explored by scholars.

Perhaps the only safe assumption to make from a macro-theory standpoint, is that when empirical evidence changes for the worse, that is, peace seemingly cedes space to conflict, realism seems vindicated. The Russian invasion of Crimea in 2014 left scholars to argue that realism was fundamentally correct even though the world had not witnessed great power competition for decades. John Mearsheimer argued convincingly for this proposition.[25] Ongoing nuclear modernization, especially if continuous and large, will likely see something similar. Nuclear modernization will most likely be interpreted as a movement away from peaceful nuclear co-existence, norms, and institutionally-driven disarmament and peace, towards a more conflict prone scenario.

Despite this vindication of the realist school of thought, taken as a single issue, nuclear modernization will be unlikely to foster a new paradigm or a renaissance of security studies as Walt argued in the early 1990s,[26] but it might offer novel theoretical work on nuclear balancing or bandwagoning. These theoretical articles, such as Waltz's[27] or Schweller's,[28] were usually written from a conventional weapons standpoint, as the nuclear proliferation debate seemed to be settled at the time. Nuclear modernization may add more detailed studies here. Further, while omni-balancing is not a novel concept,[29] the multi-modal, multidimensional balancing with uncertainty around nuclear modernization may bring about new types of exploration of balancing behavior.

Finally, while it is hard to predict with any certainty future interpretivist research on the topic, it will remain clearly a topic of interest for interpretivist scholars. For one, the power dynamic between the nuclear weapons states and the non-nuclear weapons states set forth by the Nuclear Nonproliferation Treaty may be revisited; it will be interesting to note how the hegemonic discourse on nuclear weapons use may shift with the modernization. Does the discourse around the military inapplicability or the normative stigma of nuclear weapons shift with the additional resources that are devoted to the modernization? Future discourse analysis may focus on this angle.

After the examination of macro-theory and nuclear modernization's possible effects on it, an analysis of middle-range theories follows. The critical question remains: will nuclear competition between great powers be perceived to be similar to what theory believed it to be during the Cold War? To answer this, we revisit the concept of nuclear primacy that International Relations scholarship attributed to the United States during parts of the Cold War and examine some theoretical assumptions with new evidence that has surfaced since then. By most accounts, sometime in the past fifteen years a new state entered the post-Cold War great power competition, the People's Republic of China. We ignore China's role in International Relations theory at our peril, thus, we examine whether some conventional theory derived from Cold War U.S.-Soviet interactions would hold in a U.S.-China relationship in the near future.

Rethinking Cold War nuclear deterrence in light of new evidence: MAD, nuclear primacy, and the role of the U.S. nuclear arsenal today

New research that looks at the impact of technology on counterforce missions has been bringing to light new information about the Cold War deterrence dynamics between the U.S. and the Soviet Union. The new argument is that MAD or mutual vulnerability was not a stable or static condition, but in reality much more fluid and subject to political considerations and technological capabilities situation. Groundbreaking research couched in rigorous archival work pictures MAD not as the inevitable cage of mutual vulnerability both sides are trapped in – as the conventional International Relations has it – but as a malleable place where possibilities existed of escaping it. From the 1980s onwards, technological advancements in intelligence gathering as well as nuclear targeting made mutual vulnerability questionable and nuclear superiority a realistic objective.[30] In more detail, it has been argued that the primary goal of Cold War U.S. strategy was to use its strategic nuclear arsenal to control escalation once tactical nuclear weapons were used in Europe in defense of allies.[31] U.S. conventional inferiority pushed U.S. strategists towards thinking of ways of unpunished use of tactical nukes against the Soviet Union. By holding hostage critical nodes of the Soviet nuclear arsenal – in essence threatening a pre-emptive strike that could decapitate and hence deprive the Soviet Union of a credible retaliatory strike – the U.S. sought to solve its infamous Cold War dilemma that asked it to 'sacrifice LA for Berlin'.[32]

Put differently, recent research makes the claim that the U.S. during the Cold War period never accepted mutual vulnerability vis-à-vis the Soviet Union and drastically sought ways to escape MAD, aspiring at enhancing the credibility of its nuclear umbrella.[33] The development of new and successful intelligence collection techniques worked towards this direction. Unknown until now, scholarly research reveals the possibility that the U.S. had sufficient information that could have credibly threatened the Soviet Union with a decapitation strike.[34] In other words, this evidence suggests that the defense-dominated strategic environment of the Cold War defined by mutual vulnerability was in fact an offense-dominated one, which explains massive nuclear arsenals disproving arguments of bureaucratic pathologies. Examined under this light, accumulation of warheads and delivery systems is viewed as a rational behavior of the two sides that were caught in a trap of an action-reaction spiral and in pursuit of a credible retaliatory strike capability against their adversary.

The idea of using intelligence that promises the destruction of Soviet second-strike capabilities impacted heavily the strategic thinking at the time; albeit differently in the think tank and scholarly communities compared to the military circles. For strategists at RAND, war fighting doctrines and damage limitation ideas sprung from the need to avoid total genocides and to make even nuclear war a more rational and controlled type of endeavor.[35] Counterforce targeting instead of counter-value, which would also target important locations in a country other than just military targets, became the objective which was enabled by promising intelligence.[36] The idea was a 'turning the screw' type of strategy that would hold hostage more and more counterforce sites which could be targeted and eventually hit based on negotiations and the adversary's behavior – compromise or not.

For the military, counterforce and decapitation – with a pre-emptive blow rather than damage limitation through a 'turning the screw' strategy – was needed because they deemed that U.S. command and control (C2) centers were not capable of absorbing a first strike before retaliating.[37] In fact, it was exactly this piece of information that separated the think tank world from the military when the former asked for a calibrated counterforce strategy, and the latter advocated for an all-out counterforce approach; both enabled by intelligence collection and the conviction that the Soviet Union hidden nuclear arsenal could be pinpointed and targeted. What the brilliant nuclear strategists in the think tank and academia did not take into account was the striking vulnerability of the U.S. C2 centers to the extent that the U.S. second strike capability was doubted after a Soviet nuclear surprise attack.[38] It was this piece of information on the vulnerability of U.S. communications that convinced U.S. military leaders that the best shot they had was to engage in an all-out nuclear war, where the whole Soviet arsenal needed to be destroyed in one massive U.S. nuclear strike.

Counterforce strategies were embraced by both communities for another reason; the highly assertive nature of the Soviet regime over its military which convinced U.S. strategists of the low prospects of pre-delegation of the use of nukes to regional commanders. Based on recent evidence, Soviet strategists were aware of their vulnerability and constantly sought to establish mutual vulnerability with the U.S. through a range of options from redundancy of C2 posts, to the conception, development, and deployment of a doomsday machine, the Perimeter system, widely known as the 'Dead Hand'.[39] The Dead Hand would be activated by a number of factors such as seismic or thermal waves sent in the event of a nuclear attack. If the Soviet leadership was decapitated, Perimeter would send a number of missiles that would overfly nuclear sites and activate delivery systems mated to nuclear warheads.[40] The existence of Perimeter became known recently, paradoxically, because the Soviets were scared of a U.S. counter-innovation destined to neutralize its effect.

Recent evidence and a more careful look at Cold War deterrence makes one think that MAD was more of a rhetorical device driven by political reasons rather than an unquestionable reality that defined the relationship between the two dominant nuclear arsenals. Fear from both sides for a pre-emptive decapitation strike led to pre-delegation policies both in the U.S. and the Soviet Union, as mentioned in the previous paragraph. For the U.S., recent evidence suggests that the concept of absolute presidential control over the nuclear button is an over-simplified idea that only scratches the surface of the empirical record.[41] In reality, both the U.S. and the Soviet Union were deeply worried about the survivability of their C2 centers which directly questioned their capacity of a second strike, let alone of a carefully calibrated nuclear warfighting plan.

That said, the examination of the Cold War nuclear balance asks for a more nuanced approach that reconsiders the sources of mutual vulnerability. New research openly questions the explanatory power and empirical veracity of the existential deterrence theoretical wave showing that two nuclear states are not pre-determined to enter a defense-dominated system. Nevertheless, one can argue that the Cold War system was, in fact, dominated by mutual vulnerability; C2 mutual vulnerability rather than second strike mutual vulnerability. C2 vulnerability is highly destabilizing given that it gives the advantage to the attacker and creates first strike incentives if one is sure that its first strike will incapacitate the enemy's retaliatory strike by disrupting, and not necessarily destroying its retaliatory mechanism. Therefore, the Cold War was far less stable than many scholars perceive. On the contrary, it consisted of an endless action-reaction chain where the Soviet Union sought to establish its nuclear punishment strike in a credible way against U.S. actions that kept holding it hostage.

New evidence on the Cold War interaction between the two main nuclear powers and the pillars of deterrence promise to help us revise our traditional thinking and re-formulate middle range theories that are more empirically validated thanks to newly de-classified archives. At the same time, understanding the past helps us ask better questions about the future. Based on the recent scholarly work, U.S. modernization can be examined through the light of newly offered scholarly arguments on the U.S. concerns and fears for the survivability of its C2 centers, along with its propensity towards counterforce as a way to control escalation and reinforce the credibility of its nuclear umbrella. The Cold War environment has changed dramatically but some of its features have persisted. The U.S. still needs to reassure its allies of its commitment – in East Asia, especially, given the rise of China – and one cannot help but wonder to what extent the U.S. feels more confident about the survivability of its C2 centers today given the rise of other domains of war such as cyber warfare. Modernization choices will probably be highly affected by the two just-mentioned questions while at the same time keeping an eye on another competitor's behavior. U.S. choices to deal with China, either by downplaying or emphasizing the political relevance and/or military utility of nuclear weapons in the relationship, will potentially impact our future theorizing of deterrence and 'compellence' in great power relations given China's distinctly different approach to nuclear weapons.

Post-Cold War nuclear deterrence: Chinese and American attitudes towards the bomb and its deterrent effect

China shares some common characteristics as a nuclear power with the former Soviet Union. The Chinese Military Commission claims the absolute control over the use of nuclear weapons and pre-delegation of the use of nuclear weapons to regional nuclear complex commanders to secure the survivability of its second strike is highly improbable.[42] At the same time, China has a remarkably smaller arsenal than the former Soviet Union and a well-established intention not to engage in a nuclear arms race with the U.S.[43] Furthermore, the Chinese understanding on the utility of nuclear weapons and, subsequently, any strategic and operational concepts about their use, is fundamentally different from the American one, which asks for caution when one applies Cold War concepts and examples in the context of the U.S.-China nuclear relationship. For instance, Chinese experts have been quite vocal about the role of their nuclear arsenal assigning it one unique role: counter-coercion.[44] Strategic ambiguity that surrounds the Chinese views on escalation, however, does not allow much space for certainty with the following question: Does China assign its nuclear deterrent a dual role against both conventional and nuclear attacks or both, which makes it an advocate of existential deterrence? Or does China feel that its nuclear deterrence allows it to conduct military operations at lower levels, an idea based on the assumption of escalation being controllable given the threat of nuclear use, which alludes to the stability-instability paradox? These are questions that have not been clarified yet and demonstrate the differences in the Chinese way of thinking when it comes to the strategic and political value of nuclear weapons, as well as their operationalizations in a real conflict scenario.

That said, Chinese differentiations in their understanding of the use of the nuclear weapon directly question the universality of Cold War middle-range theories and ask for analyses that open up the black box of the state in an effort to understand unit level characteristics that lead to variation in nuclear postures.[45] As it is explained below, China has been very modest and careful with the modernization of its nuclear arsenal which comes as a stark contrast with modernization in its conventional arsenal. Scholars have tried to understand the reasons behind this variation, approaching the question through a variety of ideational and/or material variables.[46] Nevertheless, there is no agreement in the policy or academic world with regard to China's strategic objectives behind a restrained development of its nuclear arsenal. Some pundits have wondered if China was moving towards a limited deterrence, while others view Chinese drivers of its nuclear policy to be the increasing conventional U.S. capabilities that threaten the Chinese retaliatory strike.[47] Understanding the drivers behind Chinese choices on its nuclear strategy is of critical importance given that they can shed light on its thinking behind nuclear weapons' utility and contribution to both political maneuvering and war-fighting. In the following section and based on open source evidence, we broadly delineate some differences in the way China and the U.S. perceive deterrence and, therefore, the potential impact that might have on middle-range theorizing.

U.S. views on deterrence: a show of force equals willingness to fight?

Well-established theories of deterrence usually break the concept down into two parameters: 1) capability and 2) willingness to implement the threat.[48] Deterrence theories often operationalize the elusive concept of credibility based on the two just-mentioned variables. Whereas operationalizing and measuring capabilities might not prove to be a challenging task, the same cannot be said about willingness to proceed with escalation. Even though analytically the concepts might appear to be distinctly separated, in reality, they are inter-related, especially in the minds of U.S. policy makers. To elaborate more, the U.S. has tried hard to solve its alliance dilemmas and reliability questions by looking for technologies that minimizes cost in the event of U.S. involvement. Since the adversary's deterrent strategy might be based on the asymmetry of interest between itself and the U.S. in the event of a conflict – a point that is further elaborated in the next section – that might involve allies or take place away from the U.S. homeland, the purpose is for the adversary to raise the cost for the U.S. as high as possible for the Americans to think that the cost exceeds the interest that is at stake.[49] In these cases, U.S. reliability as an ally is decreased. A solution to this conundrum for the U.S. is to invest in capabilities that help with cost minimization such as the Conventional Prompt Global Strike (CPGS) or Ballistic Missile Defense (BMD). Put differently, it would be safe for one to assume that the U.S. rebalances the asymmetry of interests by investing in new technologies that decrease the cost of its intervention in remote areas where its global interests are at stake, but not the security of the homeland.

That said, the U.S. conceptualization of deterrence puts the emphasis on capabilities and regards them as the source of willingness to proceed with the materialization of threats which is also reflected in its recent modernization plans.[50] It is these capabilities that are expected to signal the U.S. commitment to be involved in the event of a crisis to both allies and adversaries. Deterrence, however, is in the eye of the beholder. To what extent the United States' potential adversaries' cost-benefit calculations are influenced by the U.S. conceptualization of deterrence through capabilities that mirror – or not – willingness, is a question to be examined further. In other words, one can be far from sure that his/her deterrent is reinforced by the introduction of cost-minimizing capabilities in a situation perceived by the other side as one defined by asymmetry of interests, as we show below. A more thorough look into the other sides' calculations, views on the usability of nuclear weapons, and understandings of the balance of both capabilities and interests, is of critical importance.

China's views on deterrence: uncertainty and strategic ambiguity

China, on the other hand, has a long history of regarding the bomb as politically powerful but *militarily irrelevant.* At the same time, China has invested heavily in developing an undersea deterrence through the building of the Xia class U.S. ballistic missile submarines (SSBNs) and later the more capable Jin class.[51] The Jin class SSBN can fire SLBMs JL-2, the upgraded version of JL-1 that was carried by the Xia class SSBNs.[52] The main objective behind the undersea deterrence was to achieve greater quietness for the submarines and more assured penetrability for the SSBN launched missiles.[53] The Chinese, however, are somewhat insecure on their second-strike capability – even after the successful development of the new Jin-class submarine – predominantly based on the vulnerability of its submarines against U.S. Anti-Submarine Warfare (ASW).[54] The concern is fueled by fears that a Chinese SSBN would have a hard time traveling undetected through the Pacific until it could be well within range to fire against an American city on the West Coast (or even Guam or Hawaii for that matter).[55] More questions are raised by the decision to allow the Chinese SSBNs to patrol without nuclear warheads in their tubes, mainly due to the peculiar Chinese civil-military relations and the Party's willingness of absolute control over its arsenal. This basically means that in the event of a crisis, Chinese Jin-class submarines would have to return to their bases in order to load the warheads, which makes them even more vulnerable.

If a Chinese retaliatory strike delivered by a submarine is not guaranteed, then it is the land leg of the nuclear deterrent that provides the backbone of the Chinese second-strike capability. Through the addition of road mobile and solid fueled missiles – the DF-31 and DF-31A – China tried to increase the survivability of its arsenal, which lies at the heart of its deterrent posture. China is also expected to add a new generation missile, the DF-41, to its arsenal, according to a DoD report to Congress.[56] Admiral Haney, Commander of the United States Strategic Command (STRATCOM), stated that the Chinese are "conducting flight tests of a new mobile missile, and developing a follow-on mobile system capable of carrying multiple warheads."[57] Even though the details of the program are not known, the objective of new mobile land-based delivery systems is to maintain a limited but effective second strike capability in order to deter first strikes.[58] According to a DoD report to Congress on Chinese military capabilities, the newly added DF-41 ICBM is MIRVed which makes its interception by missile defenses an increasingly difficult task. The developments in land-based delivery systems notwithstanding, the Chinese nuclear arsenal remains numerically at the levels of the American nuclear arsenal during the 1950s.[59]

That said, the Chinese nuclear arsenal given its size and limited technological sophistication of its undersea deterrent, is still subject to speculation with regard to the certainty of its second strike capability.[60] Given new information on the feasibility of a counterforce strike, thanks to intelligence on location and critical nodes of the opponent's nuclear arsenal, one wonders how China will feel about the reliability of its own first strike. A change in its posture from retaliation once the first blow is absorbed to Launch on Warning or Launch under Attack has been speculated to be one of the potential changes reflecting Chinese fears over U.S. counterforce capabilities.[61] Nevertheless, China has not invested in offshore early warning systems which makes the detection of an incoming strike impossible. A LoW or LuA posture would require drastic changes such as mating delivery systems to warheads and increasing the readiness of the arsenal which is a journey China is far from willing to embark upon. China's stance illustrates once again its distinct behavior as far as nuclear weapons are concerned and its emphasis on the uncertainty of even one warhead surviving as a source of deterrence. In Major General Yao's words:

> For a state adopting a no-first-use policy and intending not to waste too much money on unusable weapons, dependence on opaqueness to bring about greater deterrent value is a wise choice. One can achieve deterrence through the certainty of prospective costs outweighing prospective gains, as well as through the uncertainty in cost/gain calculations. Deterrence works not only to reverse the enemy's original intention, but also to prevent him from forming such an intention for lack of information.[62]

Major General Yao's comments provide some answers to China's conceptualization of its nuclear strategy. With a big emphasis on uncertainty and manipulation of risk given the eventuality of information gaps, China assigns a big deterrent force to its nuclear arsenal; its limitations notwithstanding. Here the break from U.S. views on what produces deterrence is quite obvious. In addition, China's downplaying of the role of nuclear weapons in the context of a conflict in its periphery is further reinforced by its idea of asymmetry of interests. In a Taiwan conflict scenario, China perceives the balance of interests seriously tipping towards its side, which increases its credibility to use force making its deterrent more robust. The same applies, albeit to a lesser degree, in any conflict in its periphery. One can assume that the perceived asymmetry of interests might be at the origins of China's conviction that escalation will be contained. China regards any conflict in its periphery as a limited war scenario for the U.S. armed forces and decision makers.[63] Within this context, nuclear weapons are deemed to be irrelevant due to the non-high stakes combined to the non-proportionality of a nuclear strike that automatically renders all nuclear threats non-credible.[64] Bruce Blair argues that China has been closely observing U.S. nuclear brinkmanship cases and strongly believes that all nuclear powers do not regard nuclear threats as credible, their rhetoric notwithstanding.[65]

Of course, deterrence dynamics between the U.S. and China are quite different from the nuclear great power Cold War competition between the Soviet Union and the United States. Distinct Chinese attitudes towards the bomb deny an action-reaction, negative spiral dynamic where the technologically superior wins the coercion game and directly question nuclear weapons' military usability, which directly impacts U.S. credibility of nuclear threats. Middle range theories such as deterrence have always been closer to the empirical reality than macro-level theories given their focus on policy recommendations and problem-solving. As such, it is imperative for deterrence theorists to re-evaluate their claims and position them in the post-Cold War great power dyad. In this regard, U.S. modernization will push deterrence theorists towards rethinking nuclear coercion in a new technological environment, where new capabilities are quickly reaching the realm of concerning possibility; while at the same time, integrating unit level characteristics to make new claims on what makes a credible threat or not. Within the context of the U.S. nuclear arsenal modernization and the reviving of the Great Power competition, understanding the receiving end – China – and its own conceptualizations of what makes a threat credible is more imperative than ever in the post-Cold War environment.

Nuclear proliferation: revisiting extended theoretical debates?

This detailed examination of the intersection of nuclear policy and theory allowed us to draw some cautious conclusions about the future of nuclear competition. We also offered some insights into how both macro- and middle-range theories may evolve in light of nuclear modernization. Leaving the *terra firma* of historical analysis and analogies, we now turn to an exploratory endeavor to propose some ways in which other, non-paradigmatic, theoretical arguments may themselves be affected. On the one hand, as we have suggested earlier, a full-on renaissance of International Relations theory in response solely to the current U.S. nuclear modernization plans remains improbable. On the other hand, a renaissance of nuclear security studies is, on the balance, certainly plausible. The ongoing nuclear modernization may prompt re-envisioned theoretical pieces examining the future of nuclear competition, mimicking Herman Kahn's seminal piece,[66] or the nuclear debate modernized and revisited, in line with Waltz and Sagan's debate.[67] Thus far, from a nuclear nonproliferation policy perspective, Scott Sagan, who argued that horizontal proliferation would bring about negative consequences, has won over the policy elite, but nuclear modernization, if construed as a type of vertical proliferation, may re-ignite the debate.

As to the conceptual idea of a nuclear free world, and theoretical works underpinning this proposed meta-state, the end-goal will likely be further removed. The nuclear proliferation and nuclear stockpiles globally have followed a parabolic path, with peaks during the latter years of the Cold War. Nuclear armament limitation talks, such as SALT I and SALT II, and nuclear strategic arms reductions treaties, such as START I, II, and New START, have reinforced this path. Further, the hitherto unchallenged linear nature of nuclear drawdown will likely be revisited. In short, nuclear modernization, if it brings with it a perceived increase in proliferation as well, may challenge both this parabolic assumption and the linearity of nuclear drawdown that has happened since the peak point. We are bound to witness articles and theoretical exploratory pieces on the validity of these assumptions, producing critical questions and invigorating debates.

In addition to nuclear proliferation-centered theoretical work, three sets of extended and expanded theoretical arguments that are only indirectly linked to nuclear modernization warrant examination. Nuclear questions may precipitate into: alliances, especially alliance cohesion issues; U.S. foreign policy decision-making theories; and international organizations and global governance questions, including treaty negotiation theories. As these are indirect potential theory-expanding effects, only a brief survey of the most likely potential ramifications on these theories is provided.

One concrete effect of the nuclear modernization is the development of the new U.S. non-strategic nuclear weapon, the B61-12;[68] currently in use by NATO's nuclear sharing,[69] and already posing a number of theoretically interesting issues. First, the conceptual validity of a 'new weapon' will be tested later in the decade by the B61-12 but is currently under debate with the W76. The Trump administration has argued that the weapon is not a new one,[70] but some theorists are bound to ardently disagree. Theoretical arguments, concept development, and concept stretching in the Sartorian sense are sure to abound,[71] both on the pages of International Relations journals and in policy papers, as this topic is of contestation in the political arena.

Further, since some of these non-strategic nuclear weapons are hosted by European NATO member states, those European nations who host U.S. nuclear weapons will be facing domestic political and electoral discussions. We expect theory to examine this question in detail in the near future as well. While alliance cohesion work usually suggests that the presence of elite consensus is sufficient for military decisions to be upheld by governments, as detailed by Sarah Kreps,[72] the nuclear question is much more political. There is a reason the nuclear posture of the North Atlantic Alliance, the defense and deterrence posture review or DDPR for short, has not had a review for quite a while – no one wants to re-open this debate as its resolution is far from certain.[73] Nuclear modernization may force this chapter to be reopened, and theoretical work will surely offer more elaborate views on this.

From a U.S. foreign policy decision-making standpoint, the decision to begin nuclear modernization will provide great empirical testing grounds for novel theoretical essays. First, it will offer a new test case for the rational actor model and the bureaucratic politics model,[74] and second, it will offer avenues for theory development, given President Trump's non-traditional political leadership style.[75] In addition, the question of who wins the fight over "separated institutions sharing power" as political scientist Neustadt famously quipped,[76] will re-emerge, and both popular foreign affairs articles, such as one's that challenged President Obama's legacy,[77] and analytical ones, will likely be published.

Finally, since the nuclear proliferation agenda is global, the debates around the issue will provide new theoretical contributions for security governance and international organizational theory as well. New theoretical work that challenges the current modus operandi of states contracting around uncertainty with international treaties may emerge,[78] as states may feel pressured to manage the certain uncertainty that nuclear modernization will bring with it.[79] The Intermediate-Range Nuclear Forces (INF) treaty's demise will be a significant entry into the expanded development of relevant theoretical work.

Where we can draw safer conclusions are the methods that will be employed to investigate the nuclear modernization phenomenon and how the generation of theory development will likely proceed. The two most likely types of methods that will be employed is process-tracing and case studies, specifically focusing on nuclear modernization. The field of nuclear proliferation studies and nuclear modernization has had only limited empirical evidence in the past, with currently nine nuclear weapons states and four additional ones that have had them at some point. Any new empirical evidence in such a small-n, qualitative analytical field will most likely result in interesting theory-expanding case studies. Previous work, for example, has examined peripheral cases, such as the Libyan case in countering nuclear proliferation,[80] the abdication of nuclear weapons program in South Africa,[81] and the renunciation of a nuclear weapons capability in former USSR states.[82] In such an empirically limited field, even instances where the ultimate outcome did not result in nuclear weapons, in-depth case studies and theory-building happens, for example in the case of Egypt.[83] Thus it is a safe prediction that theory-building and theory-testing case studies[84] will be written around nuclear modernization.

Conclusions and the theory's way forward

The annals of political science, the pages of journals of International Relations scholarship, and historical evidence clearly demonstrates that changes to the empirical realm, the social world, states, organizations, and individuals interact in, generate advancements in the world of theory. Theory is not insulated from empirics, and policy changes in the nuclear world, an empirically-sparse sub-field within the broader agenda of International Relations scholarship, is no exception. Thus it is unsurprising that nuclear modernization, as a set of policy changes the current U.S. administration is pursuing, is bound to have a theory-generating effect on scholarship, but predicting the exact outcomes of this theory generation is an impossible proposition. As such, certainty was not our goal in this chapter. Rather, we hoped to reveal through the exploration of possible future theoretical development some expectations we have on the vector that the field of International Relations will take in its evolution in response to nuclear modernization.

This broad and encompassing survey of the anticipated changes in the field of International Relations revealed few key findings. On the one hand, at the macro-level, where theories are deductive in nature and are foremost concerned with parsimony, we argued that nuclear modernization will not cause a paradigmatic catharsis, but will have some impact on the undulation within the ever-continuing debate between realism, liberalism, and constructivism. On the other hand, where modernization will likely present most of its theory-altering power is at the middle-theory range, as the inductive nature of middle-range theories are always closer to empirical reality at the expense of parsimony. One example of theory development we expanded on here was the proposed changing calculations about deterrence in a more complex security environment, especially given the conceptual challenges to nuclear primacy and the difference in dynamics of interaction that the U.S.-China relationship brings about, that were not the case for the better part of the Cold War.

While the conclusions drawn about future theory are derived in an analytically rigorous manner, they only are limited in nature and uncertain in specificity. Inasmuch as prediction is a possibility in the field of International Relations, it remains, very much in sync with the field itself, a probabilistic, not deterministic endeavor. What is certain is that nuclear modernization will generate new debates and alter existing theoretical frameworks. To some extent, we hope that this chapter will provide ideas for future scholars on where to look to begin their own work on theory development.

Notes

1 Robert Owen Keohane, *Neorealism and Its Critics* (New York: Columbia University Press, 1986), 52–57.
2 Bernard Brodie, *The Absolute Weapon: Atomic Power and World Order*, 1st edition (Freeport, NY: Harcourt Brace Jovanovich, Inc., 1972).
3 Albert Wohlstetter, "The Delicate Balance of Terror," *Foreign Affairs*, January 1959, www.foreignaffairs.com/articles/1959-01-01/delicate-balance-terror.
4 Robert J. Art, "Between Assured Destruction and Nuclear Victory: The Case for the 'Mad-Plus' Posture," n.d., 20; Charles L. Glaser, *Analyzing Strategic Nuclear Policy* (Princeton, NJ: Princeton University Press, 1990), 9; Robert Jervis, "Why Nuclear Superiority Doesn't Matter," *Political Science Quarterly* 94, no. 4 (1979); Richard K. Betts, *Nuclear Blackmail and Nuclear Balance* (Washington D.C.: Brookings Institution Press, 1987).
5 Kenneth N. Waltz, "Nuclear Myths and Political Realities," *American Political Science Review* 84, no. 3 (September 1990), 730–745, https://doi.org/10.2307/1962764.
6 Vipin Narang names this assertion the 'existential bias' thesis, see Vipin Narang, *Nuclear Strategy in the Modern Era: Regional Powers and International Conflict* (Princeton, NJ: Princeton University Press, 2014).
7 Michael Krepon, "The Stability-Instability Paradox, Misperception, and Escalation Control in South Asia," *Prospects for Peace in South Asia* (2003), 261–279.
8 Fred Kaplan and Martin J. Sherwin, *The Wizards of Armageddon*, 1st edition (Stanford, CA: Stanford University Press, 1991); Henry A. Kissinger, "Force and Diplomacy in the Nuclear Age," *Foreign Affairs* 34 (1956–1955), 349.
9 Herman Kahn and Thomas Schelling, *On Escalation: Metaphors and Scenarios*, 1st edition (New Brunswick, NJ: Transaction Publishers, 2009); Colin S. Gray, "Theater Nuclear Weapons: Doctrines and Postures," *World Politics* 28, no. 2 (January 1976), 300–314, https://doi.org/10.2307/2009894; Colin S. Gray, "Nuclear Strategy: The Case for a Theory of Victory," *International Security* 4, no. 1 (1979), 54–87, https://doi.org/10.2307/2626784; Colin S. Gray and Michael Howard, "Perspectives on Fighting Nuclear War," *International Security* 6, no. 1 (1981), 185–187, https://doi.org/10.2307/2538536.
10 Kingston Reif, "U.S. Nuclear Modernization Programs," *Arm Control Association*, www.armscontrol.org/factsheets/U.S.NuclearModernization

11 Keir A. Lieber and Daryl G. Press, "The End of MAD? The Nuclear Dimension of U.S. Primacy," *International Security* 30, no. 4 (2006), 7–44; Keir A. Lieber and Daryl G. Press, "The New Era of Counterforce: Technological Change and the Future of Nuclear Deterrence," *International Security* 41, no. 4 (April 2017), 9–49, https://doi.org/10.1162/ISEC_a_00273.
12 Austin Long and Brendan Rittenhouse Green, "Stalking the Secure Second Strike: Intelligence, Counterforce, and Nuclear Strategy," *Journal of Strategic Studies* 38, no. 1–2 (January 2, 2015), 38–73, https://doi.org/10.1080/01402390.2014.958150.
13 Pavel Podvig, "Russia's Current Nuclear Modernization and Arms Control," *Journal for Peace and Nuclear Disarmament* 1, no. 2 (July 3, 2018), 256–267, https://doi.org/10.1080/25751654.2018.1526629.
14 Paul J. Smith, "China-Japan Relations and the Future Geopolitics of East Asia," *Asian Affairs: An American Review* 35, no. 4 (January 2009), 230–256, https://doi.org/10.3200/AAFS.35.4.230-256.
15 Charles L. Glaser and Steve Fetter, "Should the United States Reject MAD? Damage Limitation and U.S. Nuclear Strategy Toward China," *International Security* 41, no. 1 (July 2016), 49–98, https://doi.org/10.1162/ISEC_a_00248.
16 Hans J. Morgenthau, *Politics Among Nations: The Struggle for Power and Peace*, 5th edition, Revised (New York: Alfred A. Knopf, 1978).
17 Keith Krause and Michael C. Williams, "Broadening the Agenda of Security Studies," *Mershon International Studies Review* 40 (1996), 229–254.
18 David Baldwin, "The Concept of Security," *Review of International Studies* 23, no. 1 (April 2001), 5–26.
19 Stephen M. Walt, "International Relations: One World, Many Theories," *Foreign Policy*, No. 110, Special Edition: Frontiers of Knowledge (Spring 1998), 29–32, 34–46.
20 Daniel Maliniak, Amy Oakes, Susan Peterson, and Michael Tierney, "International Relations in the U.S. Academy," *International Studies Quarterly* 55, no. 2 (2011), 437–464.
21 Nina Tannenwald, "The Nuclear Taboo: The United States and the Normative Basis of Nuclear Non-Use," *International Organization* 53, no. 3 (Summer 1999), 433–468.
22 Proliferation Security Initiative. State Department website, www.state.gov/t/isn/c10390.htm
23 The Wassenaar Arrangement homepage, www.wassenaar.org/
24 Zangger Committee homepage, http://zanggercommittee.org/
25 John Mearsheimer, "Why the Ukraine Crisis Is the West's Fault," *Foreign Affairs*, September–October 2014 Issue.
26 Stephen Walt, "The Renaissance of Security Studies," *International Studies Quarterly* 35, no. 2 (1991), 211–239.
27 Kenneth N. Waltz, "The Origins of War in Neorealist Theory," *The Journal of Interdisciplinary History* 18, no. 4, The Origin and Prevention of Major Wars (Spring 1988).
28 Randall L. Schweller, "Bandwagoning for Profit: Bringing the Revisionist State Back in," *International Security* 19, no. 1 (Summer 1994), 72–107.
29 Steven R. David, "Explaining Third World Alignment," *World Politics* 43, no. 2 (January 1991), 233–256.
30 Lieber and Press, "The End of MAD?"; Lieber and Press, "The New Era of Counterforce"; Brendan R. Green and Austin Long, "The MAD Who Wasn't There: Soviet Reactions to the Late Cold War Nuclear Balance," *Security Studies* 26, no. 4 (October 2, 2017), 606–641, https://doi.org/10.1080/09636412.2017.1331639; Long and Green, "Stalking the Secure Second Strike."
31 Kaplan and Sherwin, *The Wizards of Armageddon*, 185–201.
32 Ibid.
33 10/1/2019 10:30:00 AM.
34 Long and Green, "Stalking the Secure Second Strike"; Lieber and Press, "The End of MAD?"; Lieber and Press, "The New Era of Counterforce"; Green and Long, "The MAD Who Wasn't There."

35 Kaplan and Sherwin, *The Wizards of Armageddon*, 74–85.
36 Ibid., 217.
37 Bruce Blair, *Strategic Command and Control*, 1st edition (Washington, DC: Brookings Institution Press, 1985).
38 Daniel Ellsberg, *The Doomsday Machine: Confessions of a Nuclear War Planner* (New York: Bloomsbury USA, 2017).
39 Blair, *Strategic Command and Control.*
40 Ellsberg, *The Doomsday Machine.*
41 Ibid.
42 Fiona S. Cunningham and M. Taylor Fravel, "Assuring Assured Retaliation: China's Nuclear Posture and U.S.-China Strategic Stability," *International Security* 40, no. 2 (October 2015), 7–50, https://doi.org/10.1162/ISEC_a_00215; Narang, *Nuclear Strategy in the Modern Era.*
43 Hans M. Kristensen and R.S. Norris, "Chinese Nuclear Forces, 2015," *Bulletin of the Atomic Scientists* 71, no. 4 (July 1, 2015), 77–84, https://doi.org/10.1177/0096340215591247. Also Yao in Twomey.
44 Wu Riqiang, "Certainty of Uncertainty: Nuclear Strategy with Chinese Characteristics," *Journal of Strategic Studies* 36, no. 4 (August 1, 2013), 579–614, https://doi.org/10.1080/01402390.2013.772510.
45 Narang, *Nuclear Strategy in the Modern Era*; M. Taylor Fravel and Evan S. Medeiros, "China's Search for Assured Retaliation: The Evolution of Chinese Nuclear Strategy and Force Structure," *International Security* 35, no. 2 (2010), 48–87.
46 Nicola Horsburgh, "Change and Innovation in Chinese Nuclear Weapons Strategy," *China Information* 26, no. 2 (2012), 185–204; Alastair Iain Johnston, "China's New 'Old Thinking': The Concept of Limited Deterrence," *International Security* 20, no. 3 (1995), 5, https://doi.org/10.2307/2539138; Jeffrey Lewis and Institut français des relations internationales, *China's Nuclear Idiosycrasies and Their Challenges*, 2013, www.ifri.org/downloads/pp47lewis.pdf; Jeannie Johnson, Kerry Kartchner, and Jeffrey Larsen (eds.), *Strategic Culture and Weapons of Mass Destruction: Culturally Based Insights into Comparative National Security Policymaking* (New York: Palgrave Macmillan, 2008); Jeffrey Lewis, *The Minimum Means of Reprisal: China's Search for Security in the Nuclear Age* (Cambridge, MA: The MIT Press, 2007).
47 For the debate see Glaser and Fetter, "Should the United States Reject MAD?; "China's Nuclear Force: Modernizing from Behind (2018)," *Union of Concerned Scientists*, accessed June 24, 2019, www.ucsusa.org/nuclear-weapons/us-china-relations/nuclear-modernization.
48 Thomas C. Schelling, *Arms and Influence* (New Haven: Yale University Press, 1966); Lawrence Freedman (ed.), *Strategic Coercion: Concepts and Cases* (Oxford and New York: Oxford University Press, 1998).
49 William Norris, "U.S. – China Escalation Dynamics," Draft Working Paper Presented at the Institute for Security and Conflict Studies, The Elliott School of International Affairs, The George Washington University, February 2019.
50 Amy F. Woolf, "Conventional Prompt Global Strike and Long-Range Ballistic Missiles: Background and Issues," Library of Congress, Washington, DC, 2019; "Special Report: *Nuclear Posture Review* – 2018," U.S. Department of Defense, accessed June 24, 2019, www.defense.gov/News/Special-Reports/NPR-2018; "The 2019 Missile Defense Review: A Good Start," accessed June 24, 2019, www.csis.org/analysis/2019-missile-defense-review-good-start.
51 Andew Erickson, Lyle Goldstein, William Murray, and Andrew Wilson (eds.), *China's Future Nuclear Submarine Force* (Maryland: Naval Institute Press, April 30, 2012).
52 Wu Riqiang, "Survivability of China's Sea-Based Nuclear Forces," *Science & Global Security* 19, no. 2 (May 2011), 91–120, https://doi.org/10.1080/08929882.2011.586312.
53 For the first patrol of the Jin-class submarine, see Jeremy Page, "Deep Threat: China's Submarines Add Nuclear-Strike Capability, Altering Strategic Deterrence," *Wall Street*

Journal, October 24, 2014, accessed May 27, 2016, http://online.wsj.com/articles/chinas-submarine-fleet-adds-nuclear-strike-capability-altering-strategic-balance-undersea-1414164738.

54 Wu Riqiang, "Survivability of China's Sea-Based Nuclear Forces," *Science & Global Security* 19, no. 2, (2011), 91–120.

55 Riqiang, "Survivability of China's Sea-Based Nuclear Forces."

56 Office of Secretary of Defense, "Annual Report to Congress: Military and Security Developments Involving the People's Republic of China 2015," Washington, DC, 2016, 25 and Cecil Haney, Testimony Before the House Armed Services Subcommittee on Strategic Forces (July 14, 2016).

57 Ibid.

58 Baohui Zhang, "U.S. Missile Defence and China's Nuclear Posture: Changing Dynamics of an Offence – Defence Arms Race," *International Affairs* 87, no. 3 (2011), 555–569.

59 "China's Nuclear Force: Modernizing from Behind (2018)," *Union of Concerned Scientists*, accessed June 15, 2019, www.ucsusa.org/nuclear-weapons/us-china-relations/nuclear-modernization.

60 For the debate on the survivability of the Chinese second strike see: Glaser and Fetter, "Should the United States Reject MAD?"; Elbridge A. Colby and Abraham M. Denmark, *Nuclear Weapons and U.S.-China Relations: A Way Forward* (Washington, DC: Center for Strategic and International Studies, March 2013).

61 Glaser and Fetter, "Should the United States Reject MAD?"

62 Yao in Christopher P. Twomey (ed.), *Perspectives on Sino-American Strategic Nuclear Issues* (New York: Palgrave Macmillan, 2008). Emphasis by underlining given by the authors.

63 Fiona Cunningham, "Maximizing Leverage: Explaining China's Strategic Force Postures in Limited Wars," PhD dissertation, Massachusetts Institute of Technology, Cambridge, 2018.

64 Todd S. Sechser and Matthew Fuhrmann, *Nuclear Weapons and Coercive Diplomacy* (Cambridge: Cambridge University Press, 2017); Cunningham and Fravel, "Assuring Assured Retaliation"; Caitlin Talmadge, "Would China Go Nuclear? Assessing the Risk of Chinese Nuclear Escalation in a Conventional War with the United States," *International Security* 41, no. 4 (April 2017), 50–92, https://doi.org/10.1162/ISEC_a_00274.

65 Bruce G. Blair and Chan Yali, "The Fallacy of Nuclear Primacy," *China Security*, Autumn, 2006, 27.

66 Herman Kahn, *Thinking About the Unthinkable in the 1980's* (London: Touchstone, 1985).

67 Scott Sagan and Kenneth Waltz, *The Spread of Nuclear Weapons: An Enduring Debate*, 3rd edition (London: W. W. Norton & Company), 2012.

68 "B61-12 Nuclear Bomb," *Air Force Technology*, www.airforce-technology.com/projects/b61-12-nuclear-bomb/.

69 "NATO's Nuclear Deterrence Policy and Forces," *North Atlantic Treaty Organization*, www.nato.int/cps/em/natohq/topics_50068.htm.

70 Paul Sonne, "Trump Poised to Get New Low-Yield Nuclear Weapons," *The Washington Post*, June 13, 2018, www.washingtonpost.com/world/national-security/trump-poised-to-get-new-low-yield-nuclear-weapons/2018/06/13/161b1466-6dac-11e8-9ab5-d31a80fd1a05_story.html?utm_term=.ea92d63eb53f.

71 Giovanni Sartori, "Concept Misformation in Comparative Politics," *The American Political Science Review* 64, no. 4 (December 1970), 1033–1053.

72 Sarah Kreps, "Elite Consensus as a Determinant of Alliance Cohesion," *Foreign Policy Analysis* 6 (2010), 191–215.

73 "Deterrence and Defence Posture Review," North Atlantic Treaty Organization, www.nato.int/cps/en/natolive/official_texts_87597.htm.

74 Graham Allison and Philip Zelikow, *The Essence of Decision* (Chicago: Pearson, 1999).
75 James Goldgeier and Elizabeth Saunders, "The Unconstrained Presidency," *Foreign Affairs*, September–October 2018 Issue, www.foreignaffairs.com/articles/2018-08-13/unconstrained-presidency.
76 Richard Neustadt, *Presidential Power: The Politics of Leadership* (New York: Mass Market, 1964), 12.
77 Zachary Roth, "Global Zero: Obama's Distant Goal of a Nuclear-Free World," *The Atlantic*, September 29, 2011.
78 Barbara Koremenos, "Contracting Around International Uncertainty," *American Political Science Review* 99, no. 4 (November 2005), 549–565.
79 Joseph Cirincione, "Addressing Proliferation Through Multilateral Agreement: Success and Failure in the Nonproliferation Regime," in Janne E. Nolan et al. (eds.), *Ultimate Security: Combatting Weapons of Mass Destruction* (Washington, DC: The Century Foundation, 2003).
80 Robert Joseph, *Countering WMD: The Libyan Experience* (Fairfax: National Institute Press, 2009), 1–24.
81 Peter Liberman, "The Rise and Fall of the South African Bomb," *International Security* (Fall 2001), 45–86.
82 William Potter, *The Politics of Nuclear Renunciation: The Cases of Belarus, Kazakhstan, and Ukraine*, Occasional Paper No. 22 (Washington, DC: Stimson Center, April 1995), 8–9, 19–26.
83 Robert Einhorn, "Egypt: Frustrated, but Still on a Non-Nuclear Course," in Kurt M. Campbell, Robert Einhorn, and Mitchell Reiss (eds.), *The Nuclear Tipping Point: Why States Reconsider Their Nuclear Choices* (Washington, DC: Brookings Institution, 2004), 43–82.
84 Alexander George and Andrew Bennett, *Case Studies and Theory Development*, 4th Printing edition (Boston: The MIT Press, 2005).

9 Modernization a determent to international security

Kingston Reif and Alicia Sanders-Zakre

Introduction

In a little-noticed comment before his controversial July 2018 summit meeting with Russian President Vladimir Putin in Helsinki, U.S. President Donald Trump characterized his government's multi-hundred billion dollar plans to replace the aging U.S. nuclear arsenal as "a very, very bad policy."[1] He seemed to express some hope that Russia and the United States, which together possess over 90 percent of the planet's nuclear warheads, could chart a different path and avert renewed military and nuclear competition. The Helsinki meeting, however, did not produce any agreement on nuclear weapons and Trump has so far shown little interest in a different nuclear spending path. Quite the opposite, in fact. "We have far more money than anybody else by far," Trump said in October 2018. "We'll build [the U.S. nuclear arsenal] up until" other nuclear-armed states such as Russia and China "come to their senses."[2]

Citing a deteriorating international security environment, the Trump administration is pursuing a significant expansion of the role and capability of the U.S. nuclear arsenal. In addition to continuing full speed ahead with its predecessor's plans to replace the nuclear triad and its associated warheads on largely a like-for-like basis, the administration is proposing to broaden the circumstances under which the United States would consider the first use of nuclear weapons, develop two new sea-based, low-yield nuclear options, and lay the groundwork to grow the size of the arsenal. At the same time, key U.S.-Russian nuclear arms control agreements, which serve to regulate the nuclear balance and prevent unconstrained nuclear competition, are now in serious doubt. The Trump administration formally withdrew the United States from the landmark 1987 Intermediate-Range Nuclear Forces (INF) Treaty in August 2019, and he has shown little interest in extending the 2010 New Strategic Arms Reduction Treaty (New START).[3] New START expires in 2021 but can be extended by up to five years subject to the agreement of Washington and Moscow. In short, the Trump administration is preparing to compete in a new nuclear arms race while simultaneously increasing the likelihood of such a contest.

The projected cost of the proposed nuclear spending spree is staggering, and it is growing. The United States currently plans to spend nearly $500 billion dollars, after including the effects of inflation, to maintain and replace its nuclear arsenal over the next decade, according to a January 2019 Congressional Budget Office (CBO) report.[4] This is an increase of nearly $100 billion, or about 23 percent, above the already enormous projected cost as of the end of the Obama administration. Over the next 30 years, the price tag is likely to top $1.5 trillion and could even approach $2 trillion.[5] These big nuclear bills are coming due as the Defense Department is seeking to replace large portions of its conventional forces and internal and external fiscal pressures are likely to limit the growth of military spending. "We're going to have enormous pressure on reducing the debt which means that defense spending – I'd like to tell you it's going to keep going up – [but] I'm not terribly optimistic," Alan Shaffer, deputy undersecretary of defense for acquisition and sustainment, said in February 2019.[6]

This chapter outlines the ways in which the Trump administration's nuclear strategy is unnecessary, unsustainable, and unsafe. It describes three realistic options to reduce spending on nuclear weapons while still maintaining a devastating nuclear deterrent. Scores of billions of dollars could be saved or redirected to higher priorities by eliminating, delaying, or scaling back the administration's proposals for new delivery systems, warheads, and infrastructure. Over the past several years, the U.S. Congress has largely supported both the Obama and Trump administration's proposals to replace the arsenal, though not without controversy. That approach can, should, and likely, will change.

The United States maintains a larger and more diverse nuclear arsenal than is required to deter and respond to a nuclear attack against itself or its allies. Despite claims that nuclear weapons "don't actually cost that much," the simple fact is that unless the administration and its successors find a 'pot of gold at the end of the rainbow', planned spending to maintain and replace the arsenal will pose a significant affordability problem, and threaten other national security priorities.[7] Moreover, the plans appear likely to increase the risks of miscalculation and unintended escalation, undermine strategic stability, and accelerate global nuclear competition.

Key leaders in Congress are increasingly aware and concerned about the rising price tag, the Trump administration's controversial proposals for expanded nuclear capabilities, and the risk of a total breakdown of the U.S.-Russian arms control architecture. These and other factors have brought far greater scrutiny to the U.S. nuclear recapitalization programs, their rationale, their cost, and policy alternatives.[8] Now is the time to re-evaluate nuclear weapons spending plans before the largest investments are made. The choice then is between the current strategy, which is excessive and unnecessary, puts the United States on course for a budgetary train wreck, and could increase nuclear risk, or a more realistic and affordable approach that still leaves the United States with a devastating nuclear force that is more than capable of deterring any nuclear threats to the United States and its allies.

Unnecessary, unsustainable, and unsafe excess

According to the Trump administration's *Nuclear Posture Review* published in February 2018, the world is a far more dangerous place than it was at the time the

Obama administration conducted its NPR in 2010.[9] "[G]lobal threat conditions have worsened markedly since the most recent 2010 NPR, including increasingly explicit nuclear threats from potential adversaries," states the 2018 review, citing Russia, China, North Korea, and Iran. "The United States now faces a more diverse and advanced nuclear-threat environment than ever before." It is true that the international security environment is less favorable than it was a decade ago. Some of the other nuclear-armed states have not been responsible actors. Technology is advancing in new and unpredictable ways. And the existing U.S. nuclear arsenal – much of which was originally built during the Cold War-era and refurbished since – is aging.

But the NPR does not provide any conclusive or compelling evidence that these challenges will be addressed or overcome by the review's strategy. The review seeks to add new capabilities and infrastructure to an arsenal that was already excessively large and redundant, and it aims to expand the circumstances under which the United States might consider the first use of nuclear weapons. In addition, the administration is undermining key arms control and nonproliferation guardrails at a time when efforts to reduce global nuclear risks are under significant stress.[10]

Taken together, these changes in policy are unnecessary, set the stage for an even greater and more unsustainable rate of spending on U.S. nuclear weapons, and increase nuclear instability in the years ahead.

Unnecessary excess

A larger arsenal than required for deterrence

The U.S. military stockpile of approximately 3,800 nuclear warheads, though far smaller than during the Cold War, is larger than is necessary to deter a nuclear attack on the United States or its allies from Russia's similarly sized nuclear arsenal, or from China, which has no more than 300 total weapons.[11] This oversized arsenal is irrelevant to the most pressing security challenges the United States and its allies face in the 21st century, including cyber threats, weak and failing states, climate change, and aggressive Russian and Chinese regional behavior. President Obama, with the support of the Pentagon, determined in 2013 that the deployed force could be reduced by up to one-third below the New START levels. Nonetheless, his administration's nuclear recapitalization plans were based on maintaining roughly the New START levels in perpetuity. The Trump administration has yet to take a position on whether to seek an extension of New START and indicated in its NPR that it does not believe further reductions in the arsenal are prudent given the security environment.[12]

But the fact remains that both the United States and Russia maintain more nuclear weapons than they need for their security. Small numerical advantages by either side would not change the fundamental deterrence equation. Indeed, the United States currently possesses more strategic delivery systems and warheads than Russia, while Russia possesses more non-strategic weapons than the United States. The New START data exchange shows that the United States has

656 deployed intercontinental ballistic missiles (ICBMs), submarine-launched ballistic missiles (SLBMs), and heavy bombers, while Russia has 524.[13] Such a disparity provides Russia with an incentive to put multiple warheads, or multiple independent reentry vehicles (MIRVs), on deployed strategic delivery systems to keep up with the United States and to invest in heavily MIRVed new systems, such as the under-development Sarmat (RS-28) heavy ICBM. Russia is believed to maintain approximately 2,000 non-strategic warheads. It is not clear how many of these weapons are readily available for offensive use. Most are in central storage and are likely dedicated as much if not more to China than the North Atlantic Treaty Organization (NATO).[14] The United States has a few hundred low-yield warheads for short-range delivery. Past U.S. reductions in the number of tactical nuclear weapons have not been conditioned on Russian reciprocity.

Ideally, the United States and Russia would agree to extend New START by a period of five years, as allowed under the treaty, and begin talks on further reductions that also address obstacles that have stymied progress in the past, such as missile defense and the nuclear arsenals of other nations. A follow-on to New START could also set limits on tactical nuclear weapons and U.S. and Russian intermediate-range ballistic missiles and ground-launched cruise missiles once prohibited by the INF Treaty. But even if such talks did begin, which appears unlikely in the near term, these talks could last years.

In the meantime, Washington should not give Moscow veto power over the appropriate size and composition of U.S. nuclear forces. Nor should it give Moscow an easy excuse to maintain a similarly bloated arsenal aimed at the United States and its allies. A decision to reduce to 1,000 deployed strategic warheads would put the United States in a stronger position to pressure Russia to rethink some of its expensive nuclear recapitalization projects and reduce its deployed strategic nuclear warheads. Perhaps more intriguingly, a U.S. willingness to reduce its arsenal could lead China to take a less passive approach to nuclear disarmament and more openly discuss the size, composition, and operations of its nuclear forces.[15]

While U.S.-Russian relations are currently strained, the decisions the United States is making now about rebuilding the nuclear arsenal are decisions that will be with it for decades to come. Decisions about force needs must consider the longer term and must weigh the opportunity costs.

A number of objections are often raised against further reducing the U.S. nuclear arsenal. One is that such cuts would reduce the U.S. ability to target adversary nuclear forces in an attempt to limit, or even preclude, the threat these forces could pose to the United States and its allies. But the United States does not need to place such a large emphasis on "counterforce" to maintain a credible deterrent. Even if the United States wanted to limit the damage Russian or Chinese nuclear forces could cause, it could not meaningfully do so without inviting a devastating nuclear response. By targeting primarily adversary leadership and war-supporting industrial targets, the United States could still hold at-risk assets valued by adversaries, reduce the number of nuclear weapons, and lessen reliance on prompt nuclear strikes.[16] In any event, a U.S. force of 1,000 deployed warheads would still provide significant counterforce capabilities.

Another objection is that further cuts would be a signal of weakness in the face of a more confrontational Russia and assertive China. But this is not a reason to maintain a nuclear force in excess of U.S. security requirements. If Washington and Moscow are not deterred by 1,000 deployed nuclear weapons deployed on multiple types of delivery systems, what logic presumes 1,550 would make a difference? In the case of China, even after dropping to 1,000 deployed strategic warheads, the United States would still enjoy a 10 to 1 advantage.

Some critics also claim that further U.S. nuclear force reductions would drive allies that depend on the U.S. nuclear "umbrella" to either capitulate to U.S. adversaries or reconsider their non-nuclear weapon status and seek their own arsenals. Such concerns merit closer inspection given the retaliatory potential of even 1,000 deployed strategic nuclear weapons, as well as the maintenance of superior U.S. conventional forces. Moreover, for a non-nuclear weapon state, such as South Korea or Japan, to openly build a nuclear arsenal would be a dramatic renunciation of its commitment not to do so under the Treaty on the Nonproliferation of Nuclear Weapons (NPT). The political costs of such a decision would be huge and likely provide fewer benefits than continuing to rely on U.S. security commitments. Furthermore, rather than express opposition to further nuclear force reductions, many U.S. allies in Europe and Asia have not only repeatedly called on the United States and Russia to extend New START, but also to achieve even deeper reductions below the limits established by the treaty.

Reassurance has always been a function of capabilities and commitment. Allies and partners are understandably concerned about the threats a more aggressive Russia and China pose to their security. These concerns are being exacerbated by President Trump's repeated assaults on the value of the U.S.-led alliance system and uncertainty in key allied capitals about what U.S. policy actually is on important foreign policy issues.[17] But the concerns of allies cannot be ameliorated by placing greater emphasis on nuclear threats and weapons.[18] The United States can continue to assure its allies and partners as it reduces its nuclear arsenal, maintains second-to-none conventional forces, and, most importantly, strengthens political relationships through reaffirmations of the value of alliances, stronger economic and cultural ties, and stepped-up dialogue.

The flawed case for new low-yield weapons

The shortcomings in the Trump *Nuclear Posture Review* (NPR) rationale for the development of additional low-yield nuclear options in the form of a low-yield SLBM warhead and sea-launched cruise missile (SLCM) are too numerous to count. The claim that Russia has lowered the threshold for the first use of nuclear weapons is hotly disputed.[19] While Russia appears to rely more heavily on nuclear weapons for its security than the United States due to its overall conventional inferiority and concerns about U.S. missile defenses, Russia's official nuclear doctrine does not support the claim that it has adopted an "escalate to win" doctrine. However, even if Moscow has done so, this is likely a result of Moscow's perceived conventional inferiority.

Regardless, adding a third and fourth low-yield warhead option to the U.S. arsenal is a solution in search of a problem. "I'm very comfortable today with [the] flexibility of our response options," Gen. John Hyten, commander of U.S. Strategic Command, said in March 2017 as the 2018 NPR was getting underway. "Our plans now are very flexible."[20] The United States already possesses hundreds of low-yield warheads as part of the air-leg of the triad and plans to invest over $150 billion in then-year dollars in the coming decades to ensure these warheads can penetrate the most advanced air defenses. This investment includes the purchase of an upgraded low-yield B61 gravity bomb, a new fleet of stealthy air-launched cruise missiles armed with refurbished low-yield capable W80 warheads, a new fleet of stealthy strategic bombers (the B-21), and a new fleet of stealthy dual-capable F-35A fighter aircraft. If current and planned air-launched options cannot deter or respond to Russian limited nuclear use, why are taxpayers being asked to spend scores of billions of dollars on these systems?

Inexplicably, the NPR fails to cite an intelligence assessment demonstrating that Russia might believe the United States would be self-deterred from using the weapons in its current arsenal (including higher-yield nuclear or conventional weapons) in response to a limited Russian nuclear attack. As John Gower, a retired rear admiral from the British Royal Navy, has written, the argument that high-yield nuclear weapons lack credibility as a deterrent against limited use, though seductive, is ultimately deceptive.[21] "It is not necessary that an adversary must be 100 percent certain you will respond as you indicate," he notes, "but the unacceptable nature of the damage he risks incurring means that he must be 100 percent certain you will not retaliate before he decides to break the taboo."

If Russian President Putin were to take the momentous decision to cross the nuclear threshold first – on a limited basis or otherwise – it would likely be because he perceives the survival of the Russian state to be at risk, or he believes Russia has a greater stake in the conflict or crisis that precipitates such use, perhaps due to divisions among NATO allies. Contrary to the NPR, which stated that Russia might contemplate using nuclear weapons first at "lower levels of conflict," Lt. Gen. Robert Ashley, director of the Defense Intelligence Agency, told the Senate Select Committee on Intelligence in January 2019 that a Russian decision to use nuclear weapons first would be driven by "the threshold they think the Kremlin would be at risk."[22] Additional U.S. low-yield nuclear options are unlikely to be effective in the face of these motivations. In fact, what is more likely to convince Putin that he could get away with limited first use are statements by President Trump questioning the value of NATO and other U.S. alliances.

Other arguments made in support of the necessity of additional low-yield weapons also miss the mark. A low-yield SLBM is not necessary to promptly strike time-perishable targets. If military action has already started in the European theater and Russia uses a low-yield nuclear weapon to seek to end a conflict it believes NATO would win conventionally, it is likely that the United States would have had sufficient time to forward deploy forces, including conventional and nuclear fighters and bombers, to provide a prompt response. Regardless, it's far from clear why the United States would need or want to respond to Russian limited nuclear use in minutes, rather than hours or even days.

The claim that a new SLCM is necessary to provide an assured theater strike option and serve as a hedge against Russian or Chinese advances in anti-submarine warfare capabilities is unconvincing. The United States is already planning to invest scores of billions of dollars in the B-21, long-range standoff weapon (LRSO), and F-35A to address the air defense challenge. A new SLCM would make it more difficult for an adversary to eliminate U.S. sea-based nuclear forces in the event of a major, unforeseen breakthrough in anti-submarine warfare capabilities. But ICBMs and bombers exist in part to guard against such a scenario. Meanwhile, the Navy is unlikely to be pleased with the additional operational and financial burdens that would come with re-nuclearizing the surface or attack submarine fleet. Arming attack submarines with nuclear SLCMs would also reduce the number of conventional Tomahawk SLCMs each submarine could carry. In other words, a new SLCM would be a costly hedge on a hedge.

Ultimately, attempting to mimic Russia by developing more low-yield options would play into Moscow's hands, since it can match NATO in the nuclear sphere. The main deterrence challenge Russia poses to the alliance is not nuclear. That means the United States should continue to invest in maintaining its overall conventional edge, buttress defenses as needed on NATO's eastern flank where Russia has local conventional superiority, and more effectively defend against and respond to Russia's use of disinformation, propaganda, and cyber tools to undermine western democratic institutions.

Redundancy within the Obama-era recapitalization program

The nuclear recapitalization plan that the Trump administration inherited from the Obama administration already included excessive amounts of redundancy. For example, the Defense Department argues that replacing the current air-launched cruise missile (ALCM) with the LRSO will extend the range of strategic bombers, ensure bombers can penetrate enemy airspace as adversaries enhance and expand their air defense capabilities, and allow individual bombers to strike more than one target with nuclear weapons at once. But it is important to remember that the United States first fielded a nuclear ALCM in the early 1980s at a time when the country did not have stealth bombers or advanced conventional cruise missiles. This is not the case today.

The range of America's existing strategic bombers is being extended by increasingly advanced long-range conventionally-armed air-launched cruise missiles. The planned introduction of at least 100 B-21 bombers, which will be armed with the upgraded low-yield B61-12 gravity bomb, conventionally armed cruise missiles such as the extended-range Joint Air-to-Surface Standoff Missile (JASSM-ER), and electronic warfare capabilities for air defense suppression, will further enhance the range and flexibility of the bomber leg of the triad.[23] Together these improvements will make the bomber leg much more formidable than it is today.

Still, some experts worry that attempting to drop a nuclear gravity bomb over a heavily defended target is too risky and might not succeed. They argue that if the United States ever used a nuclear weapon, the most prudent and least escalatory option would be to fire a nuclear-armed cruise missile from a safer standoff distance. If this concern is to be believed, then the United States should buy the LRSO instead of the B61-12. But nearly $4 billion has already been sunk into the B61-12 to date, or over half of the current projected cost of the program. Moreover, if the Air Force believes the stealth capabilities of the B-21 could be compromised soon after it is deployed, then it is reasonable to question the service's strategy for buying the bomber in the first place. For its part, U.S. Strategic Command does not appear concerned about the long-term survivability of the B-21. As Gen. Hyten told Congress in July 2017, "It's not the survivability of the bombers, it's the ability of the bombers to access targets."[24] By this Hyten means that whereas bombers armed with the B61 can only attack one target at a time, the LRSO provides each bomber the ability to attack multiple targets at one time.

It is not surprising that military planners would want many different ways of attacking a target. But the weapons associated with the other two legs of the nuclear triad – SLBMs and ICBMs – can penetrate air defenses and strike targets anywhere on the planet with high confidence. The United States possesses more warheads for these missiles than does Russia and could upload hundreds of warheads to its deployed ballistic missiles and bombers. In addition, the Navy's sea-launched Tomahawk cruise missile is also a highly capable and continually improving conventional standoff weapon, and it has an even longer range than the JASSM-ER.[25]

The sea-based leg of the triad is generally considered to be the most important leg due to the invulnerability of ballistic missile submarines underneath the ocean, the accuracy and promptness of SLBMs, and the fact that a single submarine, which currently can carry as many as 160 thermonuclear warheads, is capable of inflicting unacceptable damage on an adversary. Roughly 70 percent of U.S. accountable warheads under New START are fielded on Ohio-class submarines.

ICBMs, however, do not provide unique nuclear strike capabilities not already provided by other legs of the strategic triad. For example, a 1993 report by the Government Accountability Office (GAO) found "no operationally meaningful difference in time to target" between the ICBMs and SLBMs.[26] Moreover, to use ICBMs against targets in China or North Korea, they would have to overfly Russia. This targeting inflexibility problem greatly diminishes the utility of ICBMs outside a nuclear conflict with Russia, since overflying Russia to attack other states risks nuclear retaliation from Russia.

The main role of ICBMs today is to act as a target set – a "sponge" – that would require Russia to expend a large portion of its arsenal to try to eliminate them in the event of an all-out war, and as a hedge against an unforeseen problem with or vulnerability to the SLBM force. Though tensions between the United States and Russia have been on the rise over the past several years, the likelihood of a massive Russian surprise attack against the United States remains extremely low. Regardless, it is far from clear why maintaining 400 deployed ICBMs and purchasing a new missile with new capabilities, as opposed to continuing to rely on the existing Minuteman III missiles, is necessary to perform the sponge and hedge functions. In addition, to redundancy within the triad, the Obama administration also planned to continue the forward deployment of tactical B61 nuclear bombs in Europe, despite the fact that the military mission for which these weapons were originally intended – stopping a Soviet invasion of Western Europe because of inferior U.S. and NATO conventional forces – no longer exists. The Trump NPR augments the role of these weapons, and NATO followed suit at its July 2018 summit meeting in Brussels.[27]

When asked in 2010 if there is a military mission performed by U.S. tactical nuclear weapons in Europe that cannot be performed by either U.S. strategic nuclear or conventional forces, Gen. James Cartwright, then vice chairman of the Joint Chiefs of Staff, flatly said: "No." In fact, it is highly unlikely that the weapons could be successfully used, thereby undermining their deterrent effect. According to former Air Force General Robertus C.N. Remkes,

> Any attempt to use the B-61 will be challenged by the visibility of the many actions required to prepare the weapon and the crews for such an attack. The intended target nation of such an attack under the current planning scenarios will likely have many hours and even days to prepare its defenses and complicate matters for NATO target planners.[28]

Given their nearly non-existent military utility, the main rationale for keeping U.S. nuclear weapons in Europe is as a political symbol of the U.S. commitment to NATO, particularly to the newer members that border Russia.[29] However, this justification is also weak. U.S. nuclear forces do provide assurance to NATO allies in Europe that the United States is prepared to respond by using these weapons in the event of a nuclear attack against the alliance. But the heavy lifting of the nuclear component of extended deterrence is done by central strategic forces based in the United States and under the oceans, not the estimated 150 forward-deployed tactical nuclear weapons stored in bunkers in five NATO countries in Europe.

Even then, nuclear weapons are just a small piece of a much larger assurance puzzle, the biggest pieces of which are rooted in other elements of U.S. power. A more rational approach would be to rely instead on the strategic nuclear forces of alliance members and enhance information sharing and consultations about these forces.[30] A willingness on the part of the United States to remove its nuclear weapons from Europe could incentivize Russia to share more information about its non-strategic nuclear forces and consider limitations on them.[31] At the very least, Russia would no longer be able to point to U.S. nuclear weapons in Europe as a reason to take no action on its non-strategic nuclear weapons.

Whether one supports or opposes the current policy, the complete withdrawal of U.S. nuclear weapons from Europe over time by political and financial default can't be ruled out. It remains to be seen whether the five NATO host nations will commit to spending the political capital and economic resources necessary to replace their aging dual-capable aircraft. Germany has notably yet to do so.[32]

Excess NNSA infrastructure

The Trump Administration's NPR, which sets forth an open-ended commitment to unleashing a nuclear weapon buildup whenever the United States wishes, also lacks a compelling rationale. For example, there is no need to rush to expand the U.S. capability to produce plutonium pits, the nuclear cores of warheads, since the semiautonomous National Nuclear Security Administration (NNSA) can use pits from dismantled weapons if more are needed to sustain the arsenal. Approximately 15,000 excess pits and another 5,000 in strategic reserve are already stored at the Pantex Plant near Amarillo, Texas.[33] The Energy Department announced in 2006 that studies by Lawrence Livermore and Los Alamos National Laboratories show the pits of most U.S. nuclear weapons "will have minimum lifetimes of at least 85 years," which is about twice as long as previous official estimates.[34] Plutonium pits in the existing stockpile now average around 40 years old.

Furthermore, neither the NPR nor the Stockpile Stewardship and Management Plan explain why it is necessary to develop new warheads for U.S. ballistic missiles. The final version of the energy and water bill signed by President Trump in September 2018 called on the NNSA to produce a report estimating the cost of a possible less expensive alternative to the current plan to replace the W78, such as a life extension program similar to that performed on the W76 SLBM warhead.[35] The W76 life extension program, which will complete production at the end of fiscal year 2019, extends the life of the most prevalent warhead in the U.S. stockpile for 30 years at a cost of roughly $4 billion in then-year dollars.

The subsequent NNSA report, which was delivered to Congress in December 2018, determined that a simpler life extension program for the W78 would cost about the same, roughly $8.5 to $14.5 billion in then-year dollars, as replacing the warhead with the W87-1 and not meet military requirements.[36] But the agency did not detail how it arrived at this conclusion. The W76 life extension program refurbished far more warheads than a W78 refurbishment would and the two warheads are nearly the same age. Nor does the NPR provide a rationale for sustaining the high-yield B83-1 gravity bomb. If North Korea has built new hardened or deeply buried targets, it is far from clear why these targets cannot be held at risk by other higher-yield nuclear weapons, such as W88 warheads carried by SLBMs. Moreover, it is hard to imagine a scenario in which the benefits of detonating a megaton-class warhead on the Korean peninsula would outweigh the massive human casualty and environmental impacts.

Unsustainable excess

As the costs and scope of the Obama administration's plans to recapitalize the arsenal began to grow during the administration's second term, numerous Pentagon and NNSA officials warned about the affordability and execution challenge they posed. "We're looking at that big bow wave [of nuclear weapons spending] and wondering how the heck we're going to pay for it," Brian McKeon, former principal deputy undersecretary of defense for policy, said in October 2015.

"[A]nd probably thanking our stars we won't be here to have to answer the question."[37]

The current NPR's proposals to develop new nuclear capabilities and infrastructure will exacerbate the challenge. Withdrawal from the INF Treaty and the possible demise of New START with nothing to replace it could do the same. A reckoning is coming, the result of a massive disconnect between budgetary expectations and fiscal reality.[38] The recapitalization project cannot be sustained without significant and sustained increases to defense spending – which are unlikely to be forthcoming – or cuts to other military priorities. And the White House, Pentagon, and NNSA are in denial about the challenge. Trying to recapitalize nearly the entire arsenal at roughly the same time means less money is likely to be spent on each individual modernization program, thereby increasing the time and cost required to complete each one. The absence of reasonable planning will also result in more suboptimal choices when hard decisions become inevitable.[39] The current path is an irrational and costly recipe for diverting funds from other defense programs or buying fewer new nuclear delivery systems and reducing the size of the arsenal. The longer military and political leaders continue to deny this reality, the worse off America's nuclear deterrent and armed forces will be.

The third wave of nuclear modernization: unique challenges

Compared to the first two waves of nuclear modernization spending, several factors are poised to make the third recapitalization effort more challenging. Whereas the first two waves lasted roughly a decade, the third appears likely to need twice as long to complete. This is due in part to the fact that it now takes longer to buy new weapons systems than it did in the past. Today's systems are typically more complex, and the Pentagon purchasing bureaucracy is more risk-averse.[40]

In addition, the rising cost of the nuclear mission during the third modernization wave is scheduled to overlap with large increases in projected spending to replace and augment conventional forces.[41] In addition to continuing with plans to modernize legacy conventional weapons systems, the Trump administration is also pursuing new initiatives to maintain America's dominant military position against Russia and China. The administration wants to accelerate the development of hypersonic weapons, new types of missile defenses, and a new military department focused on space.[42] Each of the services are also calling for more force structure. The Navy wants more ships, the Air Force wants more aircraft squadrons, and the Army wants more troops.[43]

While Congress approved a major increase to defense spending in fiscal year 2018 relative to the previous year, the Pentagon's own projected spending between fiscal year 2019 and 2023 merely keeps pace with inflation, which means real defense spending would flatline, not increase, in the years ahead.[44] Replacing the U.S. nuclear arsenal is not a one-, two-, or three-year project. It will require at least 15 years of sustained increased spending. The budget requirements have been steadily increasing in recent years, but the biggest bills are slated to arrive starting in the early 2020s. According to the report of the National Defense

Strategy Commission published in November 2018, which assessed the Pentagon's 2018 National Defense Strategy, "available resources are . . . insufficient to undertake essential nuclear and conventional modernization simultaneously and rectify accumulated readiness shortfalls."[45]

The Pentagon proposes to find savings by shedding weapons that do not contribute to countering Russia and China and through a process of finding efficiencies.[46] But it remains to be seen how big the funding shifts to counter Moscow and Beijing will be – to say nothing about whether Congress will approve cuts to legacy weapons systems ill-suited to great power conflict. Past efficiency proposals have rarely been realized, been too small, or only been accomplished after an upfront investment first.[47] To make matters worse, defense spending during the Cold War was under less pressure in general than it is today. The Pentagon now has to contend with new internal budgetary challenges such as rapidly rising health care and compensation costs. According to one recent analysis, "just maintaining the size of the force will likely necessitate two to three percent growth above inflation in" the military personnel and operations and maintenance budget accounts.[48] Most importantly, the overall federal fiscal outlook is grim. The latest CBO estimates project that "federal debt held by the public is projected to grow steadily, reaching 93 percent of GDP in 2029 (its highest level since just after World War II) and about 150 percent of GDP in 2049 – far higher than it has ever been."[49] This will increase pressure to slash discretionary spending, including on defense. "Our continued plunge into debt is unsustainable and represents a dire future threat to our economy and our national security," cautioned Director of National Intelligence Dan Coats in March 2018.[50]

Meanwhile, congressional caps on discretionary spending return in fiscal years 2020 and 2021.[51] Without amendment, these could force a $171 billion decline in national defense spending, or 13 percent of the total sought for those two years by the Trump administration as of its fiscal year 2019 budget request.

Additionally, future bipartisan political support for increasing nuclear weapons spending is fragile and far from assured in the future, especially with respect to the new weapons proposed by the Trump NPR. Now in the majority in the House following the 2018 midterm elections, Democrats have conducted more aggressive oversight of the administration's nuclear policy and spending proposals. Rep. Adam Smith (D-WA), the new chairman of the House Armed Services Committee, has repeatedly made it clear that he believes the United States has more nuclear weapons than it needs for its security and can realistically afford.[52]

Finally, support for replacing the nuclear arsenal inside the Pentagon could wane. In recent years, both uniformed and civilian defense officials have repeatedly stated that the nuclear modernization plan is the number-one priority among all other competing modernization necessities.[53] However, such support is not assured moving forward. Over the past 18 months, the Pentagon has rapidly reoriented its thinking toward long-term competition with Russia and China, thereby elevating the relevance of conventional modernization.[54]

At the end of the second modernization wave, budget and political pressures, as well as changes in the strategic environment following the collapse of the Soviet Union, led to significant reductions in the number of new nuclear delivery systems that were ultimately purchased. For example, the Air Force initially sought 244 B-1 bombers but ended up buying only 100, and in 1993, the B-1 stopped participating in the nuclear mission altogether. Similarly, the planned purchase of 132 B-2 bombers war was curtailed to 21. And despite plans to build 24 Ohio-class submarines, the Navy ended production after building 18 boats, four of which were subsequently converted to a conventional role.[55]

Disarmament by default

The Trump NPR acknowledges that the cost to upgrade the nuclear arsenal is "substantial," but claims the bill is affordable because the high point of spending on nuclear weapons will be no more than 6.4 percent of Pentagon spending, a lower percentage than during the Cold War. Or as former Defense Secretary James Mattis frequently stated, "We can afford survival." And yet these statements obfuscate the severity of the nuclear budget problem facing the U.S. government.

The NPR estimate curiously does not include any of the major costs NNSA must incur to upgrade nuclear warheads and their supporting infrastructure. Despite significant budget increases over the past two years, the long-term viability of NNSA's plans is highly questionable. According to a GAO report issued in April 2017, the NNSA plans Trump inherited from Obama "do not align with its budget, raising affordability concerns."[56] Former NNSA administrator Frank Klotz said in a January 2018 interview prior to the release of the Trump NPR that the agency was already "working pretty much at full capacity."[57] And former NNSA Deputy Administrator Madelyn Creedon has noted: "Historically, neither Congress, the Department of Defense, nor the Office of Management and Budget have shown an inclination to fully fund the NNSA program of record, let alone the new initiatives such as those outlined in the 2018 NPR report."[58]

The NPR estimate also does not appear to account for the potential for significant cost growth. Unanticipated cost growth is a feature of most Pentagon acquisition programs, but because the key nuclear modernization programs are so large, variances in cost estimation can have especially significant effects.[59] Nor does the review address the additional billions of dollars that would be needed in the event of a decision to keep production lines open to build additional nuclear missiles and bombers or establish additional lines to develop ground-launched, intermediate-range cruise and ballistic missiles in a world without any negotiated constraints on Russian strategic nuclear and intermediate-range forces.

Regardless, 6 percent of a budget as large as the Pentagon's is still an enormous amount of money. By comparison, the March 2013 congressionally-mandated sequester that reduced national defense spending (minus exempt military personnel accounts) was 7 percent. Military leaders and lawmakers repeatedly described the sequester as devastating.[60] The bottom line is that the current recapitalization plans are unlikely to be executable. The Trump NPR offers no plan to pay for the rising price tag to replace the triad and upgrade conventional forces. As Gen. Robert Kehler, the former head of U.S. Strategic Command, bluntly put it in November 2017, "I am skeptical that we are capable of remaining committed to a long-term project like this without basically messing with it and screwing it up."[61] Indeed, a possible, if not likely, outcome is that the current plans will collapse under their own weight, forcing reductions in U.S. nuclear forces based on fiscal and political pressure rather than on strategic decisions – but not before hundreds of millions or even billions of taxpayer dollars are squandered.

Unsafe excess

In addition to being unnecessary and unsustainable, the policies and spending plans outlined in the Trump NPR could increase the risks of unintended escalation and miscalculation, undermine strategic stability, accelerate global nuclear competition, and threaten U.S. conventional advantages.

A new technological arms race

Though the Cold War-era numerical nuclear arms race is over, the U.S. nuclear recapitalization program is part of what some experts have described as "a dynamic technological nuclear arms race."[62] Not only is this new arms race different than its Cold War predecessor, it could also be more dangerous.[63] Despite significant reductions in the overall number of nuclear weapons, all of the world's nine nuclear-armed states are, to varying degrees or another, devoting vast sums of money to replace, upgrade and, in some cases, expand the size and lethality of their nuclear arsenals and delivery systems. Past, present, and planned U.S. efforts to sustain and replace the existing arsenal have increased and will continue to increase the military capability of the weapons across key attributes such as stealth, accuracy, range, speed, hard-target kill, and yield flexibility.[64] The more capable weapons being produced as a result of this new arms race, particularly more accurate and stealthier lower-yield weapons, could lower the threshold for nuclear use in a crisis or war.

These developments are inconsistent with the obligations of the five declared nuclear-weapon states under the NPT's Article VI requirement to "pursue negotiations in good faith on effective measures relating to cessation of the nuclear arms race at an early date and to nuclear disarmament." The 2009 final report of the bipartisan Congressional Commission on the Strategic Posture of the United States observed that other "nations may not show the nuclear restraint the United States desires or support nonproliferation efforts if the nuclear weapon states take no further agreed steps to decrease their reliance on nuclear arms."[65]

To complicate matters further, technological change and advances in conventional weapons and associated doctrines for their use pose new escalatory risks, including to the nuclear level, and threaten to erode nuclear stability.[66] Russia and the United States, as well as China, are all seeking to apply such technologies as artificial intelligence, robotics, advanced conventional strike, and cyber, among others, to offensive military use. In addition, Washington and Moscow are expanding their missile defenses, an issue which helped to stymie reductions below New START levels during the Obama administration, and pursuing next-generation technologies to improve their defensive capabilities. Beijing is also developing a missile defense architecture. All three countries have demonstrated anti-satellite capabilities.

These advances will likely put new strains on strategic and crisis stability, by reducing decision and warning time, increasing the odds of arms racing in the development of these weapons and capabilities to counter them, and reducing the likelihood of further nuclear arms reduction agreements. Indeed, Russia has attributed its pursuit of several new and exotic strategic weapons systems, including nuclear-armed hypersonic glide vehicles, globe-circling, nuclear-powered cruise missiles, and very long-range nuclear torpedoes, to concerns about the open-ended and unconstrained development of U.S. missiles defenses. Russia claims that these systems wouldn't be limited by New START because they don't use ballistic flight trajectories.

A cold shoulder to arms control

Unlike the Obama administration, the Trump administration's plan to rebuild the arsenal is not accompanied by a proactive arms control and nonproliferation agenda aimed at reducing nuclear weapons risks. Arms control only gets a brief mention at the end of the 2018 NPR and it is a generally dismissive mention at that. The document passively states that "the United States will remain receptive to future arms control negotiations if conditions permit" and to negotiations that "advance U.S. and allied security, are verifiable, and enforceable." No previous nuclear arms control agreement has included enforcement measures. The review offers next to nothing in the way of proposals to address proliferation challenges, ameliorate emerging challenges to strategic stability, and pursue disarmament steps. In addition, the review expresses opposition to U.S. ratification of the Comprehensive Nuclear Test Ban Treaty (CTBT) even though the United States decided more than a quarter-century ago to halt nuclear explosive testing and there is no technical need to resume nuclear testing. No reason or justification for rejecting the goal of CTBT ratification is provided. Since the release of the NPR and the arrival of arms control skeptic John Bolton as National Security Advisor, the Trump administration has withdrawn from the 2015 Joint Comprehensive Plan of Action (the Iran nuclear deal), has withdrawn from the INF Treaty, and, so far, failed to take Russia up on its offer to begin discussions about an extension of New START and resume a regular dialogue on strategic stability.[67]

Prior to joining the Trump administration, Bolton called New START "an execrable deal" and urged Trump to abrogate the agreement.[68] Administration officials have stated that they have plenty of time to make a decision on whether to extend the treaty and will take several issues into account.[69] These include Russia's manufacturing of concerns about U.S. compliance with the treaty, whether Russia would agree to limit all of the new strategic weapons systems it is developing, and

Russia's compliance with other arms control agreements.[70] While New START appears to be in serious jeopardy, the U.S. military and intelligence community continue to stress the national security benefits of the treaty. Without the INF Treaty and New START, there would be no legally binding, verifiable limits on the U.S. and Russian nuclear arsenals for the first time since 1972. The collapse of the U.S.-Russian arms control architecture would mean Russian nuclear forces would be unconstrained, our insight into Russian nuclear force structure and modernization would be curtailed, and the incentives to engage in costly nuclear competition would be magnified.

The danger of a new low-yield weapons

In addition to being unnecessary, new low-yield weapons could increase the risk of nuclear conflict. As former White House official Lynn Rusten notes, "new low-yield nuclear weapons would not 'raise the bar' for nuclear use; they would lower it because they increase the contingencies and planning for use and fuel the illusion that a use of nuclear weapons could remain limited and not escalate into a large-scale nuclear exchange."[71] The belief that a nuclear conflict could be controlled is dangerous thinking. The fog of war is thick, the fog of nuclear war would be even thicker. Large or small, nuclear weapons are extremely blunt instruments, both in terms of their destructive power and the taboo associated with the fact they have not been used in 70 years.

Placing greater emphasis on low-yield options could also have the perverse effect of convincing Russia that it could get away with limited nuclear use without putting its survival at risk. According to Creedon, who served during the Obama administration as assistant secretary of defense for global strategic affairs before joining NNSA, "Signaling that a low-yield weapon would be used to respond to low-yield weapon use might persuade Russia to lower the nuclear threshold, thus risking nuclear war-fighting."[72]

In the case of the proposed low-yield SLBM warhead, given that U.S. strategic submarines currently carry SLBMs armed with higher-yield warheads, how would Russia know (or discriminate) that an incoming missile armed with a low-yield warhead was not actually armed with high-yield warheads? How would it know that such limited use would not be the leading edge of a massive attack, especially if the targets would not be battlefield targets but targets of high-value to the Russian leadership, as some have claimed? The answer is that Russia would not know, thereby increasing the risks of unintended escalation.

Firing a single low-yield warhead from a strategic submarine could also undermine the survivability of the most important leg of the U.S. nuclear triad, which would be at a premium in the event of a nuclear conflict. As Creedon notes:

> The sea leg of the nuclear triad is the most survivable leg in large part due to the ability of Ohio-class submarines to be invisible in the open ocean. Launching a high-value D5 missile from a ballistic missile submarine will most likely give away its location. China and Russia are expanding their

ability to detect a missile launch and will be able to locate a U.S. submarine if it launches a D5 missile. Is having a low-yield warhead worth the risk of exposing the location of a ballistic missile submarine at sea?[73]

The United States has never before armed ballistic missiles with a low-yield warhead. The proposal to do so in the NPR brings into play new scenarios for how the United States might use prompt-strike, long-range SLBMs, including against battlefield targets, which require further examination and analysis. It also could prompt Russia and China to deploy low-yield warheads on ballistic missiles, an outcome the United States should want to avoid.

During the George W. Bush administration, bipartisan majorities in Congress killed administration proposals to develop a variable yield "robust nuclear earth penetrator" and put a conventional warhead on Trident missiles citing concerns about need, usability, and unintended escalation.[74] These same concerns also apply to the low-yield SLBM and SLCM proposals.

What makes the Trump administration's proposal to develop additional low-yield nuclear weapons even more concerning is that the 2018 NPR envisions a greater role for nuclear weapons against a wider range of threats. Unlike the previous administration, the Trump administration defines the "extreme circumstances" under which the United States would consider nuclear use more broadly to include "significant non-nuclear strategic attacks" against "U.S., allied or partner civilian population or infrastructure, and attacks on U.S. or allied nuclear forces, their command and control, or warning and attack assessment capabilities." Threatening nuclear retaliation to counter new kinds of asymmetric attacks would lower the threshold for nuclear use, increase the risks of miscalculation, and make it easier for other countries to justify excessive roles for nuclear weapons in their policies. Such threats are also unlikely to be proportional and therefore would be difficult to make credible.

Nuclear cruise missiles, ICBMs, and the risks of accidental nuclear war

The plans to augment the role of nuclear-armed cruise missiles and replace the ICBM force raise additional stability concerns. Proponents of the LRSO claim that it would simply sustain an existing capability – not expand that capability. In reality, the LRSO is likely to have greatly enhanced capabilities relative to its predecessor, and will be mated to the B-52, B-2 and B-21 bombers, whereas the current ALCM can only be delivered by the B-52.[75] U.S. nuclear stealth bombers have never carried stealthy nuclear cruise missiles. The LRSO raises serious questions about stability that have yet to be fully explored. Some sources claim that the Pentagon is envisioning potential uses for the new cruise missile that go beyond "the original mission space" of the ALCM, namely in limited nuclear war-fighting contingencies involving China.[76] Furthermore, as stressed by William Perry, President Bill Clinton's defense secretary, and Andrew Weber, a former assistant secretary of defense, "cruise missiles are a uniquely destabilizing type of weapon" due to the fact that "they can be launched without warning and come in both nuclear and conventional variants."[77]

Deploying nuclear-armed SLCMs on U.S. surface ships or attack submarines would pose similar problems. Currently, the Navy only fields conventional Tomahawk SLCMs. By adding nuclear SLCMs to the mix, any use of conventional Tomahawks, especially in a conflict with another nuclear-armed state, would inherently send a nuclear signal. This would diminish the utility of the missiles and boats that carry them in a conventional conflict and increase the potential for miscalculation.

The NPR claims that the administration would consider forgoing the development of a new nuclear SLCM, which would take nearly a decade to field, if Moscow "returns to compliance with its arms control obligations, reduces its non-strategic nuclear arsenal, and corrects its other destabilizing behaviors." This requirement is so sweeping that it lacks any realistic negotiating value. Moreover, instead of compelling a change in Russian behavior for the better, a new SLCM could prompt Russia (and China) to build more intermediate-range nuclear weapons systems, including weapons on land in violation of the INF Treaty. As Rusten points out:

> Russian investments in new intermediate-range strike capabilities appear driven by perceptions of vulnerability to U.S. and NATO prompt-strike and missile defense capabilities. Compounding Russia's perceived vulnerabilities will prompt more countermeasures, not submission. By what logic should the United States fuel an incipient arms race by pursuing nuclear weapons systems it does not need?[78]

The vulnerability and risks of accidental launch associated with U.S. land-based ICBMs have long been debated. Given their vulnerability in fixed though hardened silos, the United States retains plans to launch ICBMs under attack before adversary missiles could destroy them to guard against a "disarming" first strike. This means the president might have only three to six minutes to decide how to respond after an incoming attack is detected. Though the risk of accidental launch is low, early warning systems have in the past experienced false alarms and some experts are increasingly worried that a third-party cyber-attack could trigger a false alarm.[79]

Meanwhile, the Air Force is planning to replace the Minuteman III ICBM with a more capable and accurate missile to overcome advancing adversary defensive measures. Gen. Robin Rand, the former commander of Air Force Global Strike Command, told Congress in 2016 that:

> Improved ICBM capability and accuracy has the benefit of providing ICBM strike planners the weaponeering options of either achieving a higher probability of effect on a given target; using fewer warheads per target while still achieving the desired level of effect and thus allowing more targets covered; or provide opportunities to potentially reduce yield size while still achieving the desired level of effect.[80]

This suggests the United States is seeking to improve the counterforce warfighting capabilities of the ICBM force. Supporters of retaining and recapitalizing the ICBM leg of the triad argue that eliminating ICBMs would drastically reduce the number of U.S. targets an adversary would need to destroy in a disarming first strike from over 500 to less than ten.[81] While the United States would still retain SSBNs at sea in the event of such an attack, ICBMs advocates claim that it would be unwise to rely on the invulnerability of submarines in perpetuity given advances in possible detection technologies.

But some former government officials, military leaders, and prominent experts call for eliminating ICBMs due to their lack of a unique mission and the risk they could trigger an accidental nuclear war. These voices include Perry and Gen. Cartwright. "As we make decisions about which weapons to buy, we should use this simple rule," they wrote in a November 2017 op-ed advocating the elimination of ICBMs and ALCMs. "If a nuclear weapon increases the risk of accidental war and is not needed to deter an intentional attack, we should not build it."[82]

Eliminating ICBMs would also remove the targets for a large portion of Russian ICBMs. As nuclear strategist Thomas Schelling put it in 1987:

> If we unilaterally dismantled our land-based missiles, we would instantly deprive a large part of the Soviet land-based missile force for its raison d'être. It might look to them as if they had much less to preempt. They actually would not, because the U.S. missiles they might have preempted were redundant in the first place. . . . So if we cannot dismantle their land-based missiles by negotiation, we may gain a lot by dismantling their targets instead.[83]

In addition, relinquishing ICBMs would reduce the risks in an escalating crisis or conflict that Russia might try to preemptively strike them to limit the damage to itself. According to Ryan Snyder, a nuclear physicist and previously a visiting research fellow at the Arms Control Association:

> In a severe crisis that involves missile exchanges, major battles, or the loss of some strategic nuclear targets by conventional means, however, an adversary could conclude that escalation to the nuclear level is imminent. And if they decide that it is preferable to be attacked with fewer weapons rather than more, attention could turn to the ICBMs and other vulnerable U.S. military targets.[84]

Other skeptics of the value of ICBMs note that even if ICBMs are retained, keeping a launch-under-attack policy is unnecessary and dangerous.[85] Given the size, accuracy, and diversity of U.S. forces, the remaining nuclear force would be more than sufficient to deliver a devastating blow to any nuclear aggressor. The survivability of the U.S. command and control system and sea-based leg of the triad means that U.S. leaders have time to consider how to respond to even a massive nuclear attack. No U.S. leader should be put in a situation that could lead to the use of nuclear weapons in a matter of minutes based on false information, however small the risk.

The risks of new warheads

NNSA's plans to develop new ballistic missile warheads have prompted concerns about compromising confidence in the reliability of the arsenal. The original plan for the IW-1 proposed using parts from two different existing warheads that have never been used together. A newly built W78 warhead, even if it is not interoperable, could introduce unwelcome doubts about reliability into an otherwise well-tested and reliable stockpile.[86] In addition, the NNSA's plans to expand the infrastructure for plutonium pit production could raise significant safety and environmental problems. Safety problems at Los Alamos forced the lab to stop production of plutonium pits from 2013 to 2016. Significant safety lapses in the plutonium operations at Savannah River also have been documented in recent internal government reports, according to a 2018 report by the Center for Public Integrity.[87]

Damaging opportunity costs

Prioritizing an excessive nuclear improvement program could compromise investments in conventional capabilities and other critical national security programs. In this context, it is useful to compare the looming spending binge on nuclear delivery systems and their supporting infrastructure to overall Pentagon acquisition spending, as these are the areas of the budget where dollars are likely to be most directly in competition. The CBO estimates that by the early 2030s, spending on nuclear weapons will peak at a mammoth 15 percent of the Pentagon's total acquisition costs, more than triple the current share.[88]

At a service level, the opportunity costs are particularly stark. The Navy has repeatedly warned that the projected $128 billion cost to develop and purchase 12 new Columbia-class ballistic missile submarines will devastate its shipbuilding budget.[89] Similarly, the Air Force's new ICBM program will compete with other service priorities, such as the F-35 and new tanker programs.

At the NNSA, increased spending in recent years on warhead life extension programs has led to cutbacks in funding for critical stockpile surveillance work and the Stockpile Stewardship Program, which assesses and certifies the reliability of the stockpile in the absence of nuclear testing, as well as the agency's efforts to prevent nuclear terrorism and proliferation.[90] Every dollar Washington spends to maintain a bloated nuclear arsenal is a dollar that cannot be spent on conventional military capabilities more relevant to countering Russia and China and assuring U.S. allies. It is not in the U.S. interest to engage in a tit-for-tat race with the Russians to rebuild an excessively large nuclear force, particularly if it comes at the expense of needed conventional improvements, especially programs to maintain military readiness and a technological edge with regard to Russia and China.

Conclusion

The United States is planning to spend hundreds of billions of dollars over the next two decades to rebuild a nuclear arsenal much like the one it has today and to last another 50 years. The current and planned U.S. financial investment in nuclear forces is unrivaled by any other nuclear power. But the spending plans face significant budgetary, programmatic, and political challenges. The question, then, is not whether the United States is falling behind its competitors – it is not – but whether the size and configuration of the current arsenal and the Trump recapitalization plans are necessary, sustainable, and safe. The answer is that the current course is unnecessary, unsustainable, and unsafe – and must be rethought. It is not too late to pursue a different path. Now is the time to re-evaluate nuclear weapons spending plans before the largest investments are made.

A common argument made in support of the approach proposed in the Trump NPR is that the only alternative is to allow the U.S. nuclear deterrent to waste away. But this is a false choice. The October 2017 CBO report evaluated numerous alternatives to the current sustainment and recapitalization program that, if pursued, would reduce nuclear weapons spending while still maintaining a reliable and credible nuclear deterrent. The options range from blended reductions to each or several legs of the triad to moving to a dyad. The report measured the capability of the alternatives relative to that of the current program across four metrics: the number of warheads, crisis management, limited nuclear strikes, and large-scale nuclear exchanges. Of course, pressure on the defense budget cannot be relieved solely by reducing nuclear weapons spending. A significant portion of the overall cost of nuclear weapons is fixed. Key components of the supporting infrastructure, such as the command-and-control systems and nuclear laboratories, would remain whether the United States possessed 10 nuclear weapons or 10,000. That said, changes to the nuclear replacement program could make it easier to execute and ease some of the hard choices facing the overall defense enterprise, while still leaving a force more than capable of deterring nuclear attacks against the United States or its alliance partners.

There are three realistic options for reducing U.S. spending on nuclear weapons that would save at least an estimated $29 billion to $282 billion from fiscal year 2017 to 2046. The bulk of these savings would occur over the first 20 years of the 30-year period. Unlike the Trump NPR, the second and third options in particular would reflect a nuclear strategy that reduces reliance on nuclear weapons, emphasizes stability and survivability, deemphasizes nuclear warfighting, reduces the risk of miscalculation, and is more affordable and executable.

The baseline for this estimate is the October 2017 CBO estimate of the Obama administration's plans to maintain and replace U.S. nuclear forces and their supporting infrastructure and our projection of the costs of the additions proposed by the Trump NPR.[91] With the exception of the first option, which is in fiscal year 2018 dollars, all estimates are in fiscal year 2017 dollars unless otherwise noted. The estimated savings from each option includes savings from research and development, procurement, and operations and sustainment unless otherwise noted.

The first option would eliminate the additions to the Obama-era recapitalization program proposed in the Trump NPR. This option would avoid an estimated $28.8 billion in additional costs above the CBO baseline over the next

30 years. The second option would reduce costs by more cost-effectively deploying 1,550 New START-accountable strategic warheads. This option would save an estimated $120.5 billion relative to the CBO baseline and $149.3 billion when combined with eliminating the Trump additions over the next 30 years. The third option would eliminate the ICBM leg of the triad and decrease the number of New START-accountable strategic warheads to 1,000. This option would save an estimated $253 billion relative to the CBO baseline and $281.8 billion when combined with eliminating the Trump additions over the next 30 years.

Each option would involve the purchase of new fleets of ballistic missile submarines, SLBMs, and long-range penetrating bombers. All of the options would also retain at least one low-yield nuclear delivery option. None of the options would alter current plans to upgrade nuclear command, control, communications, and early-warning capabilities. Upgrading these capabilities and reducing their vulnerability to attack should be a top priority and will likely require additional funding beyond what the Pentagon has identified to date. All of the options also assume the continued maintenance of the three national nuclear laboratories: Los Alamos, Livermore, and Sandia.[92] In fact, even under the most ambitious cost-saving option, the United States would still be poised to spend roughly $1 trillion on nuclear weapons over the next 30 years due to the large fixed costs of the nuclear weapons enterprise.

This chapter has outlined the ways in which the Trump administration's nuclear strategy is unnecessary, unsustainable, and unsafe. It describes three realistic options to reduce spending on nuclear weapons while still maintaining a devastating nuclear deterrent. Scores of billions of dollars could be saved or redirected to higher priorities by eliminating, delaying, or scaling back the administration's proposals for new delivery systems, warheads, and infrastructure.

The choice is between the current strategy, which is excessive and unnecessary, puts the United States on course for a budgetary train wreck, and would increase nuclear risk, or a more realistic and affordable approach that still leaves the United States with a devastating nuclear force that is more than capable of deterring any nuclear threats to the United States and its allies.

Given the stakes, the choice should be obvious.

Notes

1 "Remarks by President Trump and Prime Minister May of the United Kingdom in Joint Press Conference," The White House Office of the Press Secretary, July 13, 2018, www.whitehouse.gov/briefings-statements/remarks-president-trump-prime-minister-may-united-kingdom-joint-press-conference/.

2 Eli Watkins and Maegan Vazquez, "Trump Threatens Nuclear Buildup Until Other Nations 'Come to Their Senses'," October 24, 2018, www.cnn.com/2018/10/22/politics/donald-trump-russia-china-inf/index.html.

3 Press Statement from Michael R. Pompeo, "U.S. Intent to Withdraw from the INF Treaty," February 2, 2019, www.state.gov/secretary/remarks/2019/02/288722.htm.

4 U.S. Congressional Budget Office, "Projected Costs of U.S. Nuclear Forces, 2019 to 2028," January 2019, www.cbo.gov/system/files?file=2019-01/54914-Nuclear Forces pdf.

5 U.S. Congressional Budget Office, "Approaches for Managing the Costs of U.S. Nuclear Forces, 2017 to 2046," October 2017, www.cbo.gov/system/files?file=115th-congress-2017-2018/reports/53211-nuclearforces.pdf.
6 Paul Mcleary, "Bigger Debt Means Smaller Budgets (but We Need More Shipyards): DoD Officials," *Breaking Defense*, February 15, 2019, https://breakingdefense.com/2019/02/bigger-debt-means-smaller-budgets-but-we-need-more-shipyards-dod-officials/.
7 "Remarks by Deputy Secretary of Defense Carter at the Aspen Security Forum at Aspen, Colorado," July 18, 2013, http://archive.defense.gov/Transcripts/Transcript.aspx?TranscriptID=5277.
8 Leo Shane III and Joe Gould, "Military Budget Showdown Set as Democrats Advance Authorization Bill Over GOP Objections," *Defense News*, July 12, 2019, www.militarytimes.com/news/pentagon-congress/2019/07/12/military-budget-showdown-set-as-democrats-advance-authorization-bill-over-gop-objections/.
9 "Nuclear Posture Review," Office of the Secretary of Defense, February 2018, https://media.defense.gov/2018/Feb/02/2001872886/-1/-1/1/2018-NUCLEAR-POSTURE-REVIEW-FINALREPORT.PDF.
10 Kingston Reif and Kelsey Davenport, "Trump's Threat to Nuclear Order," *War on the Rocks,* October 12, 2017, https://warontherocks.com/2017/10/trumps-threat-to-nuclear-order/.
11 "Factsheet: Nuclear Weapons: Who Has What at a Glance," *Arms Control Association*, last updated June 2018, www.armscontrol.org/factsheets/Nuclearweaponswhohaswhat.
12 Kingston Reif, "No Arms Control Advances in U.S.-Russian Talks," *Arms Control Today*, September 2018, www.armscontrol.org/act/2018-09/news/arms-control-advances-us-russian-talks.
13 "Factsheet: New START Treaty Aggregate Number of Strategic Offensive Arms," U.S. Department of State, last modified July 1, 2019, www.state.gov/new-start-treaty-aggregate-numbers-of-strategic-offensive-arms-10/.
14 Hans M. Kristensen and Robert S. Norris, "Russian Nuclear Forces," *Bulletin of the Atomic Scientists*, May 4, 2018, https://thebulletin.org/2018/05/russian-nuclear-forces-2018/.
15 Kingston Reif and Victor Mizin, "A Two-Pronged Approach to Revitalizing U.S.-Russia Arms Control," The Deep Cuts Commission, July 2017, http://deepcuts.org/publications/working-papers/#c1951.
16 Bruce Blair, Jessica Sleight, and Emma Claire Foley, "The End of Nuclear Warfighting: Moving to a Deterrence-Only Posture," Program on Science and Global Security, Princeton University, September 2018, www.globalzero.org/wp-content/uploads/2018/09/anpr-ExecSummary.pdf.
17 Robin Emmott, "Trump's Portrayal of NATO in Crisis Carries Risks for Alliance," *Reuters*, July 13, 2018, www.reuters.com/article/us-nato-trump-analysis/trumps-portrayal-of-nato-in-crisis-carries-risks-for-alliance-idUSKBN1K3286.
18 Jon Wolfsthal, "America Can't Fix Its Problems with New Nukes," *Foreign Policy*, June 6, 2017, https://foreignpolicy.com/2017/06/06/america-cant-fix-its-problems-with-new-nukes/.
19 Kristin Ven Bruusgaard, "Here's Why U.S. Tactical Nukes Are a Bad Idea," *The National Interest*, December 10, 2018, https://nationalinterest.org/feature/heres-why-us-tactical-nukes-are-bad-idea-38372.
20 Gen. John Hyten, "Military Reporters and Editors Association Conference – Keynote Speech," March 31, 2017, www.stratcom.mil/Media/Speeches/Article/1153029/military-reporters-and-editors-association-conference-keynote-speech/.
21 John Gower, "The Dangerous Illogic of Twenty-First-Century Deterrence Through Planning for Nuclear Warfighting," *Carnegie Endowment for International Peace*, March 6, 2018, https://carnegieendowment.org/2018/03/06/dangerous-illogic-of-twenty-first-century-deterrence-through-planning-for-nuclear-warfighting-pub-75718.

22 U.S. Congress, Senate, Select Committee on Intelligence, "Worldwide Threats," January 29, 2019, www.intelligence.senate.gov/hearings/open-hearing-worldwide-threats.
23 The Defense Department has begun development of an "extreme range" version of the JASSM that would have an even longer range than the JASSM-ER. See John Tirpak, "Lockheed Martin Developing 'Extreme-Range' JASSM Variant," *Air Force Magazine*, September 13, 2018, www.airforcemag.com/Features/Pages/2018/September%20 2018/Lockheed-Martin-Developing-Extreme-Range-JASSM-Variant.aspx.
24 U.S. Congress, Senate, Committee on Armed Services, "United States Strategic Command Programs: Hearing Before the Senate Committee on Armed Services," April 4, 2017, www.armed-services.senate.gov/hearings/17-04-04-united-states-strategic-command-programs.
25 David Hambling, "Tomahawk Missiles Will Get Twice as Deadly by Blowing Up Their Own Fuel," *Popular Mechanics*, January 15, 2016, www.popularmechanics.com/military/weapons/news/a18994/tomahawk-missiles-fuel-air-explosion/.
26 U.S. Government Accountability Office, "The U.S. Nuclear Triad: GAO'S Evaluation of the Strategic Modernization Program," June 1993, www.gao.gov/products/GAO/T-PEMD-93-5.
27 Monica Montgomery, "NATO Presses Stand on Nuclear Weapons," *Arms Control Today*, September 2018, www.armscontrol.org/act/2018-09/news/nato-presses-stand-nuclear-weapons.
28 Steve Andreasen and Isabelle Williams, "Options for NATO Nuclear Sharing Arrangements," *Nuclear Threat Initiative*, 2011, www.nti.org/media/pdfs/NTI_Framework_full_report.pdf?_=1322694001?_=1322694001.
29 Kingston Reif, "U.S. Nukes in Europe Are Useless," *RealClearDefense*, September 4, 2014, www.realcleardefense.com/articles/2014/09/05/us_nukes_in_europe_are_useless_107414.html.
30 Steve Andreasen, "Rethinking NATO's Tactical Nuclear Weapons," *Survival*, October–November 2017, www.tandfonline.com/doi/abs/10.1080/00396338.2017.1375225.
31 James Doyle, "Better Ways to Modernise the U.S. Nuclear Arsenal," *Survival*, August–September 2016, www.tandfonline.com/doi/abs/10.1080/00396338.2016.1207946.
32 Bojan Pancevski, "In Germany, a Cold War Deal to Host U.S. Nuclear Weapons Is Now in Question," *The Wall Street Journal*, February 12, 2019, www.wsj.com/articles/in-germany-anger-at-trump-throws-cold-war-nuclear-pact-into-question-11549976449.
33 "Groups Release Key DOE Documents on Expanded Plutonium Pit Production, DOE Nuclear Weapons Plan Not Supported by Recent Congressional Actions," *Nuclear Watch New Mexico*, May 31, 2018, https://nukewatch.org/pressreleases/PR-Pit-Production-Docs-5-31-18.pdf.
34 James Sterngold, "Doubts Cast on Need for New Nukes/Study Finds Plutonium May Last Twice as Long as Expected," *SF Gate*, November 15, 2006, www.sfgate.com/news/article/Doubts-cast-on-need-for-new-nukes-Study-finds-2466709.php.
35 U.S. Congress, "The Energy and Water Development and Related Agencies Appropriations Act, 2019," September 21, 2018, www.congress.gov/115/bills/hr5895/BILLS-115hr5895enr.pdf.
36 Rachel S. Cohen, "NNSA: New GBSD Warhead Plan Costs Slightly More Than Refurbished Option," *Airforce Magazine*, January 31, 2019, https://www.airforcemag.com/nnsa-new-gbsd-warhead-plan-costs-slightly-more-than-refurbished-option/.
37 Aaron Mehta, "Is the Pentagon's Budget About to Be Nuked?" *Defense News*, February 5, 2016, www.defensenews.com/2016/02/05/is-the-pentagon-s-budget-about-to-be-nuked/.
38 Kingston Reif and Mackenzie Eaglen, "The Ticking Nuclear Budget Time Bomb," *War on the Rocks*, October 25, 2018, https://warontherocks.com/2018/10/the-ticking-nuclear-budget-time-bomb/.

39 Rick Berger, "How Much Will America Spend on Nukes in the Next Decade?" *American Enterprise Institute*, January 25, 2019, www.aei.org/publication/how-much-will-america-spend-on-nukes-in-the-next-decade/.
40 Reif and Eaglen, "The Ticking Nuclear Budget Time Bomb."
41 Todd Harrison, "Defense Modernization Plans Through the 2020s: Addressing the Bow Wave," Center for Strategic and International Studies, January 2016, https://csis-prod.s3.amazonaws.com/s3fs-public/legacy_files/files/publication/160126_Harrison_DefenseModernization_Web.pdf.
42 Aaron Mehta, "3 Thoughts on Hypersonic Weapons from the Pentagon's Technology Chief," *Defense News,* July 16, 2018, www.defensenews.com/air/2018/07/16/3-thoughts-on-hypersonic-weapons-from-the-pentagons-technology-chief/; Paul Sonne, "Pentagon Looks to Adjust Missile Defense Policy to Include Threats from Russia, China," *The Washington Post*, March 2, 2018, www.washingtonpost.com/world/national-security/pentagon-looks-to-adjust-missile-defense-policy-to-include-threats-from-russia-china/2018/03/01/2358ae22-1be5-11e8-8a2c-1a6665f59e95_story.html?utm_term=.29f3ef004a8d; "U.S. Space Force Estimated to Cost $13 Billion in First Five Years: Memo," *Reuters*, September 17, 2018, www.reuters.com/article/us-usa-military-space/us-space-force-estimated-to-cost-13-billion-in-first-five-years-memo-idUSKCN1LX2GZ.
43 David B. Larter, "U.S. Navy to Add 46 Ships in Five Years, but 355 Ships Won't Come for a Long Time," *Defense News*, February 12, 2018, www.defensenews.com/smr/federal-budget/2018/02/13/us-navy-to-add-46-ships-in-five-years-but-355-ships-is-well-over-the-horizon/; Stephen Losey, "Air Force Calls for 74 More Squadrons to Prepare for Possibility of War Against Major Power," *Air Force Times*, September 17, 2018, www.airforcetimes.com/news/your-air-force/2018/09/17/air-force-calls-for-74-more-squadrons-to-prepare-for-possibility-of-war-against-major-power/; Michelle Tan, "Chief: 'The Army Needs to Get Bigger," *Army Times*, October 8, 2017, www.armytimes.com/news/your-army/2017/10/09/chief-the-army-needs-to-get-bigger/.
44 Mackenzie Eaglen, "Defense Budget Peaks in 2019, Underfunding the National Defense Strategy," American Enterprise Institute, May 17, 2018, www.aei.org/publication/defense-budget-peaks-in-2019-underfunding-the-national-defense-strategy/.
45 National Defense Strategy Commission, *Providing for the Common Defense*, United States Institute of Peace, November 13, 2018, http://usip.org/publications/2018/11/providing-common-defense.
46 Todd Harrison and Seamus P. Daniels, "Analysis of the FY 2019 Defense Budget," Center for Strategic and International Studies, September 2018, www.csis.org/analysis/analysis-fy-2019-defense-budget.
47 Reif and Eaglen, "The Ticking Nuclear Budget Time Bomb."
48 Harrison and Daniels, "Analysis of the FY 2019 Defense Budget."
49 U.S. Congressional Budget Office, "The Budget and Economic Outlook: 2019 to 2029," January 2019, www.cbo.gov/system/files?file=2019-01/54918-Outlook.pdf.
50 Josh Gerstein, "Intelligence Official Warns Trump Administration on National Debt," *Politico*, February 13, 2018, www.politico.com/story/2018/02/13/trump-national-debt-intelligence-officials-407255.
51 Brendan W. McGarry, "The Defense Budget and the Budget Control Act: Frequently Asked Questions," *Congressional Research Service*, July 13, 2018, https://fas.org/sgp/crs/natsec/R44039.pdf.
52 "'Nothing Endangers the Planet More Than Nuclear Weapons', Kingston Reif interview with Rep. Adam Smith," *Arms Control Today*, December 2018, www.armscontrol.org/act/2018-12/features/nothing-endangers-planet-more-nuclear-weapons.
53 Aaron Mehta, "Mattis Enthusiastic on ICBMs, Tepid on Nuclear Cruise Missile," *Defense News*, January 12, 2017, www.defensenews.com/space/2017/01/12/mattis-enthusiastic-on-icbms-tepid-on-nuclear-cruise-missile/.

54 Reif and Eaglen, "The Ticking Nuclear Budget Time Bomb."
55 Lawrence Korb and Adam Mount, "Setting Priorities for Nuclear Modernization," *Center for American Progress*, February 2016, https://cdn.americanprogress.org/wp-content/uploads/2016/02/03053017/NuclearArsenal2.pdf.
56 U.S. Government Accountability Office, "Department of Energy: Continued Actions Needed to Modernize Nuclear Infrastructure and Address Management Challenges," February 6, 2018, www.gao.gov/products/GAO-18-374T.
57 Aaron Mehta, "As Trump Seeks New Nuke Options, Weapons Agency Head Warns of Capacity Overload," *Defense News*, January 23, 2018, www.defensenews.com/space/2018/01/23/as-trump-seeks-new-nuke-options-weapons-agency-head-warns-of-capacity-overload/.
58 Madelyn Creedon, "A Question of Dollars and Sense: Assessing the 2018 Nuclear Posture Review," *Arms Control Today*, March 2018, www.armscontrol.org/act/2018-03/features/question-dollars-sense-assessing-2018-nuclear-posture-review.
59 Reif and Eaglen, "The Ticking Nuclear Budget Time Bomb."
60 Howard McKeon and James Inhofe, "Forced Budget Cuts a Disaster for Military," *CNN*, February 20, 2013, www.cnn.com/2013/02/20/opinion/inhofe-mckeon-budget-cut-military/index.html.
61 Jon Harper, "Nuclear Triad: Pentagon Taking Steps to Modernize Global Strike Weapons," *National Defense Magazine*, November 3, 2017, www.nationaldefensemagazine.org/articles/2017/11/3/nuclear-triad-pentagon-taking-steps-to-modernize-global-strike-weapons.
62 Hans M. Kristensen, "Nuclear Weapons Modernization: A Threat to the NPT?" *Arms Control Today*, May 2014, https://armscontrol.org/act/2014_05/Nuclear-Weapons-Modernization-A-Threat-to-the-NPT.
63 Benjamin Zala, "How the Next Nuclear Arms Race Will Be Different from the Last One," *Bulletin of the Atomic Scientists*, January 2, 2019, https://thebulletin.org/2019/01/how-the-next-nuclear-arms-race-will-be-different-from-the-last-one/.
64 Hans M. Kristensen, "Nuclear Modernization, Enhanced Military Capabilities, and Strategic Stability," Presentation to Arms Control Association Annual Meeting, June 6, 2016, https://fas.org/nuke/guide/usa/Brief2016_ACA.pdf.
65 "America's Strategic Posture: The Final Report of the Congressional Commission on the Strategic Posture of the United States," United States Institute of Peace, 2009, www.usip.org/publications/2009/04/congressional-commission-strategic-posture-united-states.
66 Michael Klare, "The Challenges of Emerging Technologies," *Arms Control Today*, December 2018, www.armscontrol.org/act/2018-12/features/challenges-emerging-technologies.
67 Kingston Reif, "Arms Control on the Brink," *Bulletin of the Atomic Scientists*, November 1, 2018, https://thebulletin.org/2018/11/arms-control-on-the-brink/.
68 John Bolton, "Trump's New Start with Russia May Prove Better than Obama's," *The Wall Street Journal*, February 13, 2017, www.wsj.com/articles/trumps-new-start-with-russia-may-prove-better-than-obamas-1486941073.
69 Kingston Reif, "Republican Senators Back New START," *Arms Control Today*, October 2018, www.armscontrol.org/act/2018-10/news/republican-senators-back-new-start.
70 For more on Russia's concern's about U.S. compliance with New START, see Kingston Reif, "As INF Treaty Falls, New START Teeters," *Arms Control Today*, March 2019, www.armscontrol.org/act/2019-03/news/inf-treaty-falls-new-start-teeters. Regarding Russia's development of new kinds of strategic offensive arms, New START envisioned the possible development of such weapons during the period of implementation of the treaty. Article II of the treaty states: "When a Party believes that a new kind of strategic offensive arm is emerging, that Party shall have the right to raise the question of such a strategic offensive arm for consideration in the Bilateral Consultative Commission."

71 Lynn Rusten, "The Trump Administration's 'Wrong Track' Nuclear Policies," *Arms Control Today*, March 2018, www.armscontrol.org/act/2018-03/features/trump-administrations-wrong-track-nuclear-policies.
72 Creedon, "A Question of Dollars and Sense."
73 Ibid.
74 Tom Collina, "Trump's New 'Low-Yield' Nuclear Weapon: Two Bad Ideas Rolled into One," *The National Interest*, March 10, 2018, https://nationalinterest.org/blog/the-buzz/trumps-new-low-yield-nuclear-weapon-two-bad-ideas-rolled-one-24806.
75 Kingston Reif, "New Cruise Missile Capability Debated," *Arms Control Today*, January 2016, www.armscontrol.org/ACT/206_0102/News/New-Cruise-Missile-Capability-Debated.
76 Ibid.
77 William J. Perry and Andy Weber, "Mr. President, Kill the New Cruise Missile," *The Washington Post*, October 15, 2015, www.washingtonpost.com/opinions/mr-president-kill-the-new-cruise-missile/2015/10/15/e3e2807c-6ecd-11e5-9bfe-e59f5e244f92_story.html?noredirect=on&utm_term=.0f792c588c37.
78 Rusten, "The Trump Administration's 'Wrong Track' Nuclear Policies."
79 Bruce G. Blair, "Why Our Nuclear Weapons Can Be Hacked," *The New York Times*, March 14, 2017, www.nytimes.com/2017/03/14/opinion/why-our-nuclear-weapons-can-be-hacked.html.
80 U.S. Congress, House of Representatives, Committee on Armed Services, Subcommittee on Strategic Forces, "Fiscal Year 2017 Budget Request for Department of Defense Nuclear Forces," March 2, 2016, https://docs.house.gov/Committee/Calendar/ByEvent.aspx?EventID=104619.
81 Matthew Kroenig, "The Case for the U.S. ICBM Force," *Strategic Studies Quarterly* (Fall 2018), www.airuniversity.af.edu/Portals/10/SSQ/documents/Volume-12_Issue-3/Kroenig.pdf.
82 William J. Perry and James E. Cartwright, "Spending Less on Nuclear Weapons Could Actually Make Us Safer," *The Washington Post*, November 16, 2017, www.washingtonpost.com/opinions/spending-less-on-nuclear-weapons-could-actually-make-us-safer/2017/11/16/396ef0c6-ca56-11e7-aa96-54417592cf72_story.html?utm_term=.49e524a01c5f.
83 Thomas C. Schelling, "Abolition of Ballistic Missiles," *International Security* 12, no. 1 (Summer 1987), 179–183.
84 Ryan Snyder, "The Future of the ICBM Force: Should the Least Valuable Leg of the Triad Be Replaced?" Arms Control Association, March 2018, https://armscontrol.org/policy-white-papers/2018-03/future-icbm-force-should-least-valuable-leg-triad-replaced.
85 Robert Einhorn and Steven Pifer, "Meeting U.S. Deterrence Requirements," The Brookings Institution, September 2017, www.brookings.edu/research/meeting-u-s-deterrence-requirements/.
86 Tom Collina, "The Unaffordable Arsenal: Reducing the Costs of the Bloated U.S. Nuclear Stockpile," *Arms Control Association*, October 2014, www.armscontrol.org/reports/The-Unaffordable-Arsenal-Reducing-the-Costs-of-the-Bloated-U.S.-Nuclear-Stockpile.
87 "Nuclear Negligence," The Center for Public Integrity, https://apps.publicintegrity.org/nuclear-negligence/.
88 U.S. Congressional Budget Office, "Approaches for Managing the Costs of U.S. Nuclear Forces, 2017 to 2046."
89 Ronald O'Rourke, "Navy Columbia (SSBN-826) Class Ballistic Missile Submarine Program: Background and Issues for Congress," October 23, 2018, Congressional Research Service, https://fas.org/sgp/crs/weapons/R41129.pdf.
90 Creedon, "A Question of Dollars and Sense."

91 Though the October 2017 CBO report covers the period between fiscal year 2017 and 2046, fiscal years 2017 and 2018 have already concluded and fiscal year 2019 will be nearly halfway complete at the time of publication of this report. Thus, the savings from the options described in this report would be different, though not substantially so, if implemented starting in fiscal year 2019.

92 A smaller arsenal could substantially reduce the rationale for maintaining three nuclear weapons laboratories.

10 Modernization as a promoter of international security

The special role of U.S. nuclear weapons

Matthew Kroenig and Christian Trotti

Does nuclear modernization promote international security? All the major powers are currently modernizing their strategic nuclear arsenals, and scholars and policymakers are debating the effect of such modernization on international politics. Some argue that nuclear weapons deter a great power war and, therefore, nuclear modernization will contribute to international stability.[1] Others claim that modernization programs are fundamentally destabilizing; nuclear weapons are the deadliest weapons ever invented and their continued possession will eventually result in a catastrophic nuclear attack.[2]

This chapter introduces a third answer to the above question: it depends.

The modernization programs of the United States and its democratic allies at the core of the international system tend to promote international security. On the other hand, the nuclear modernization programs of revisionist autocratic powers intent on challenging the international system tend to threaten global peace and stability. Thus, not all modernization programs are created equal. It is ultimately the underlying geopolitical conditions, and not solely the capabilities themselves, which determine the extent to which modernization programs promote or diminish international security.

Since 1945, the United States has been the leader of a rules-based international order that has brought the world unprecedented levels of peace, prosperity, and freedom.[3] U.S. nuclear weapons have provided the security backbone of that order by deterring major power conflict in Europe and Asia.[4] Additionally, U.S. extended nuclear security guarantees have protected the entire free world and dissuaded many capable states from developing independent nuclear capabilities. The modernization of U.S. nuclear weapons will therefore continue to reinforce these major sources of geopolitical peace and stability.

Alternatively, Russia, China, and North Korea are opposed to major elements of the existing international system and seek in varying ways to disrupt or displace it. Given their revisionist geopolitical objectives, nuclear modernization in these countries tends to threaten international peace and stability.

Geopolitical intentions guide nuclear strategy, which in turn informs modernization programs. It is therefore essential to account for a state's position within the international system when evaluating the likely effects of nuclear modernization on international stability.

The rest of this essay proceeds in four parts. First, it explains why U.S. nuclear weapons serve as unique guarantor of international security in the postwar world order. Second, it argues that the United States must modernize its aging nuclear capabilities in order to sustain its longstanding role in promoting global security. Third, it addresses the ways in which revisionist powers seek to use their nuclear modernization programs for destabilizing ends. Fourth, it briefly considers the role of other nations' nuclear forces, including: Britain, France, India, Israel, and Pakistan. Finally, it concludes with implications for scholarship and for policy.

The U.S.-led, rules-based international order

Following the end of World War II, the United States and its victorious allies built the international system in which we reside today.[5] Like past hegemons, the United States ordered an international system to suit its interests. Unlike past hegemons, however, Washington did not seek to rule through formal empire. Rather, it sought to reproduce a variant of its rules-based domestic political system for the global system. This rules-based international order possesses multiple components. First, Washington and its allies established a wide range of international institutions, such as the United Nations, the World Bank, the North Atlantic Treaty Organization (NATO), and the Treaty on the Nonproliferation of Nuclear Weapons, to encourage multilateral solutions to a wide variety of challenges. Second, it sought to correct the protectionism of the interwar years by creating an open economic system defined by the free flow of trade and capital across borders. Third, the United States and its democratic allies have promoted freedom, human rights, and good governance abroad. Lastly, Washington employed its unmatched military might to provide the underlying stability and security which has allowed the overall system to function.

While the system is imperfect and the United States has sometimes disobeyed the rules it helped to write, the rules-based order has nevertheless produced remarkable results. Humanity has been blessed by the absence of great power war for over seventy years. Global standards of living are much higher today than seven decades ago.[6] And by the mid-2000s, more people were living under democratic governments than at any time in world history.[7]

As the world's most powerful state and the architect of this system, the United States advanced and defended this order through a variety of economic, diplomatic, and military means. An underappreciated but central pillar of this rules-based order, however, has been the U.S. nuclear arsenal. U.S. nuclear weapons have provided the backbone for the underlying security element of the global system.

U.S. nuclear weapons and great power peace

U.S. nuclear weapons contribute to global stability by deterring great power war. U.S. nuclear weapons have deterred revisionist autocratic powers, such as the Soviet Union during the Cold War and Russia, China, and North Korea today, from attacking U.S. allies in Europe and Asia.

Scholars have long appreciated the role of nuclear weapons in deterring war and contributing to the long period of great power peace. Indeed, much social science scholarship credits the nuclear revolution with transforming international politics and making large-scale war among the major powers unthinkable.[8]

What much of this literature ignores, however, is the unique role played by American nuclear weapons. Unlike other nuclear powers that seek to deter attacks on themselves only, the United States aims to deter attacks against the entire free world.[9] The United States extends deterrence to over thirty formal treaty allies, including twenty-eight members of NATO, South Korea, Japan, Australia, and arguably others. The U.S. policy of extended deterrence provides geopolitical stability in Europe and Asia, the two most important geostrategic regions on the planet and the home of all the major nuclear powers.

As a status quo power and leader of the international system, the United States has had little interest in or incentive to use military force to revise the territorial status quo in Europe or Asia. Following the end of World War II, for example, the United States' immediate goal was in setting up independent governments in defeated countries, allowing free elections, and returning home.[10]

The Soviet Union, on the other hand, had an explicit goal of spreading Marxist revolutions to other nations in Europe and around the world. It established puppet states under Moscow's control in the territories it occupied in Eastern Europe. It employed military force, in Hungary in 1956 and Czechoslovakia in 1968, in order to force these states to remain in the Soviet camp. The Soviet Union also attempted to use military coercion in a failed bid to wrest control of Berlin from the West in a series of crises during the late 1950s and 1960s. In Asia, Moscow gave the green light to North Korea to invade its Southern neighbor and initiate the Korean War.[11] These are only the major examples of Soviet militarized revisionism during the Cold War.

In the face of this threat from the Red Army, the United States remained in Europe at the behest of its allies to deter Russian invasion. In order to counter the Soviet Union's conventional superiority in Europe, the United States and NATO relied heavily on nuclear deterrence.[12] U.S. nuclear weapons and the requisite resolve to defend allies were critical to forcing Moscow's submission in the Berlin Crises and to deterring war in Europe. In addition, the United States became more deeply involved in Asia, intervening militarily in Korea and in the Taiwan Straits to prevent communist forces from invading or threatening their neighbors.[13] In sum, during the Cold War in Europe and Asia, the Soviet Union's foreign policy was largely oriented toward revising the status quo, while the United States generally employed force, including nuclear deterrence, in order to uphold it.

Similarly today, Russia, China, and North Korea express their displeasure with the territorial status quo in their respective regions.[14] Russian President Vladimir Putin refers to the collapse of the Soviet Union as "the greatest geopolitical catastrophe of the 21st century" and longs to re-establish a "greater-Russia."[15] He has been clear about his interest in re-establishing a sphere of influence in Eastern Europe and he has already invaded two of his neighbors, Georgia and Ukraine. China has also declared a long-term intention to redraw the map of Asia to incorporate Taiwan and it has begun taking contested territory in the South China Sea through military coercion and gray-zone tactics. North Korea has engaged in repeated bouts of aggression against South Korea and Seoul's American ally over the decades. It initiated the Korean War and seized the U.S.S. Pueblo during the Cold War. In recent years, Pyongyang has sunk a South Korean warship and shelled a South Korean island.

U.S. democratic allies in Europe and Asia consistently call upon the United States to maintain a robust nuclear strategy and posture in order to deter possible aggression from these autocratic and revisionist powers. For example, when U.S. President Obama considered the adoption of a nuclear No First Use Policy at the end of his term, U.S. allies resisted.[16] Largely in response to allied demands, the United States agreed to keep the nuclear option on the table to deter conventional aggression. It has also developed nuclear capabilities, including the supplemental, low-yield capabilities called for in the 2018 U.S. *Nuclear Posture Review*, with the explicit goal of extending deterrence and reassuring allies.[17]

Therefore, it is true that nuclear weapons have deterred great power war for over seventy years, but it is important to consider whose potential aggression has been deterred. If nuclear weapons were to be un-invented or nuclear deterrence to fail, any ensuing aggression would most likely be conducted, not by the United States, but by Moscow, Beijing, or Pyongyang against their regional neighbors. It is specifically U.S. nuclear weapons that are the most important element for keeping the peace in Europe and Asia.

U.S. nuclear weapons and nonproliferation

The United States has also contributed to the security of the rules-based international system by halting the spread of the world's most dangerous weapons. Once again, in this domain, U.S. nuclear weapons have been central to these efforts.

As the most powerful state in the international system and the leader of the rules-based order, the United States possesses a strong interest in preventing other nations from acquiring nuclear weapons.[18] Washington played a unique role in establishing the major institutions of the nuclear nonproliferation regime, including the Treaty on the Nonproliferation of Nuclear Weapons (NPT).[19] Some theorists, like Kenneth Waltz, argue that horizontal proliferation yields greater systemic stability and that Washington could benefit from slow and selective nuclear proliferation, but there are many risks to the spread of nuclear weapons, especially for the world system's leader.[20] Accordingly, this "proliferation optimism" argument has never found favor in the corridors of power.[21]

To support its goal of preventing the spread of nuclear weapons to allied states, the United States has provided these states with extended deterrence. Washington struck a bargain with allies: refrain from building your own nuclear weapons and we will protect you with U.S. nuclear weapons. As stated above, the United States currently extends its nuclear umbrella to over thirty formal treaty allies and arguably others. Without these efforts at extending deterrence to assure allies, it is likely that at least several other U.S. allies would possess independent nuclear capabilities today. U.S. assurance efforts were essential to eliminating nuclear weapons ambitions in West Germany in the early days of the Cold War, and South Korea and Taiwan in the 1970s.[22] Studies employing statistical analysis have shown that, "States receiving security guarantees from nuclear-armed superpower allies are only 22 percent as likely to explore nuclear weapons as those who do not, 13 percent as likely to pursue them, and 15 percent as likely to acquire them in a given year, conditional on their not having done so previously."[23]

To make this extended deterrent guarantee more credible, Washington sometimes goes so far as to deploy U.S. nuclear weapons on the territory of its allies. This forward-deployed American nuclear power has been referred to as a wedding ring to its allies, providing a visible symbol of America's security commitment.[24] An empirical study on the link between forward-deployed nuclear weapons and nonproliferation found that forward-deployed nuclear weapons have a perfect track record in preventing nuclear proliferation. Never has a country with U.S. nuclear weapons on its soil developed independent nuclear capabilities.[25]

To be sure, U.S. attempts at nonproliferation have occasionally failed. In the early stages of the Cold War, U.S. allies Britain and France chose to build independent nuclear arsenals, despite Washington's preferences to the contrary. Still, the small number of nuclear-armed states today is a fantastic achievement relative to pessimistic proliferation expectations of the mid-twentieth century. And it is very likely that there would be many more nuclear powers in the world today without U.S. efforts at extended deterrence and assurance.

U.S. nuclear security guarantees remain critical in the current security environment to stop the spread of nuclear weapons. With newfound questions about America's commitment to the security of its allies, there are renewed discussions in Berlin about whether Germany should consider building its own nuclear weapons.[26] A majority of the public in South Korea support Seoul going nuclear under uncertain conditions.[27] And analysts predict that a weakening of the U.S. nuclear umbrella could lead to proliferation among U.S. allies, including Japan, Poland, and perhaps other states.[28]

In contrast, other nuclear-armed powers have played a less central role in preventing the spread of nuclear weapons. Indeed, these revisionist and autocratic states have occasionally gone so far as to use their nuclear capabilities to intentionally export the bomb to other states, fueling the spread of nuclear weapons.[29] For example, the Soviet Union aided China in developing a nuclear program from 1958 to 1960 to increase the power of the communist bloc relative to the United States. Russia considered exporting a uranium enrichment plant to Iran in 1995 before Washington intervened and pressured them to cancel the transaction. China aided Pakistan due to mutual rivalry with India. Beijing also transferred sensitive nuclear technology to Iran, Algeria, and other countries as well. While Pakistani scientist Abdul Qadeer Khan created a proliferation network which enabled North Korea, Iran, and Libya to acquire sensitive nuclear technology as part of constructing an alliance of "strategic defiance" against the United States and its allies.[30] In recent years, Pyongyang has become a nuclear exporter, helping Syria to build a heavy-water reactor before it was destroyed in an Israeli air strike in 2007. All of these countries remain at risk of exporting sensitive nuclear technology in the future, intentionally fueling proliferation in a bid to advance their revisionist aims.

Some have argued that by possessing a robust nuclear arsenal, the United States instigates the spread of nuclear weapons to additional countries, but this is incorrect. They aver that the continued maintenance of an American nuclear arsenal sets a dangerous precedent, undermines the NPT, and encourages other countries to build nuclear weapons. These arguments are plausible at a superficial level, but closer investigation reveals that they are unfounded.[31] States build nuclear weapons for a wide range of security, economic, and normative reasons, but aping the behavior of the United States is not among them. Moreover, there is no correlation between the size of the U.S. nuclear arsenal and the probability that other states explore, pursue, or acquire nuclear weapons.[32]

In sum, U.S. nuclear weapons also contribute to the security of the rules-based system by reinforcing the nonproliferation regime and preventing the spread of nuclear weapons.

The imperative of U.S. nuclear modernization

United States' nuclear weapons have provided for international security and stability for the past three-quarters of a century, but, if they are to continue to play that role in the future, then they must be modernized.

The U.S. strategic nuclear forces consist of a triad of delivery vehicles: 1) submarine-launched ballistic missiles (SLBMs) deployed on ballistic missile submarines (SSBNs); 2) intercontinental ballistic missiles (ICBMs); and 3) strategic bombers (including the B-2 and B-52) carrying gravity bombs and air-launched cruise missiles (ALCMs).

These capabilities are aging, however, and must be updated. The OHIO-class SSBNs entered service in the 1980s and 1990s and were originally designed for a 30-year service life. Their service life has already been drawn out to 42 years and further extension is impossible.[33] There are only so many times a submarine can withstand the intense pressures of submerging and resurfacing, and relying on these old boats puts the life of the crew at risk.

For the land-based leg, the Minuteman III ICBMs were first deployed in 1970, and they were meant to have a 10-year service life. This service life has been extended multiple times, but the United States cannot continue to rely on this capability beyond 2030.[34]

Turning to the air leg of the triad, U.S. ALCMs are now 25 years past their expected design life.[35] The United States also needs to ensure that its stealth bombers can continue to penetrate current and future enemy air defense systems.

Beyond the delivery systems, the underlying nuclear command, control, and communications (NC3) system is also three decades old and is vulnerable to new space and cyber threats.[36] Lastly, the warheads themselves are aging, as many were built in 1976, 1978, 1980, and 1988. These warheads require life extension programs for continued survival.[37]

In sum, U.S. nuclear weapons were built at the end of the Cold War in the 1970s and 1980s. If you drive a car, it is likely that it was not built in the 1970s or 1980s. And, if it was, then it is not likely very reliable. As long as the United States possesses a nuclear deterrent, it must ensure that it is safe, effective, and reliable. This can only be achieved with modernized capabilities.[38]

In 2010, then-U.S. President Barack Obama laid out a plan to modernize America's aging nuclear arsenal over the next thirty years. He called for modernizing each leg of the nuclear triad. In 2031, the new Columbia-class SSBN will begin to replace the Ohio fleet, while the Department of Defense will research a potential replacement to the D5 SLBM.[39] The Department of Defense has also launched the Ground-Based Strategic Deterrent (GBSD) program to modernize 450 ICBM launch facilities and field 400 new ICBMs to replace the Minuteman III, beginning in 2029.[40] Meanwhile, in order to improve the survivability of the air-based leg in the face of sophisticated enemy air defenses, the U.S. will replace the B-2A with the new B-21 Raider stealth bomber in the mid-2020s. It is also replacing the ALCM with the long-range standoff (LRSO) weapon, and it is modernizing the B61 gravity bomb.[41] Simultaneously, the Department of Defense is planning NC3 advancements for improved command and control, as well as coordination with the Department of Energy to extend the service life of the warheads.[42]

President Trump decided to continue with Obama's modernization plans in his *Nuclear Posture Review*, released in 2018.[43] In addition, the Trump administration called for capability "supplements" to deal with the threat of Russian nuclear "de-escalation" strikes.[44] These low-yield nuclear weapons include installing a low-yield warhead on a small number of SLBMs and the resurrection of a nuclear sea-launched cruise missile.

These modernization plans have strong bipartisan support with senior defense officials on both sides of the aisle attesting to the need for these capabilities.[45] Nevertheless, some analysts have questioned the desirability and feasibility of U.S. nuclear modernization.[46] Some have argued that U.S. nuclear modernization is too expensive. The Congressional Budget Office (CBO) estimates that the cost of nuclear modernization over the next thirty years will come to over $1 trillion. Critics argue that this is unaffordable and this money could be better spent on other priorities. While $1 trillion is certainly a large number, to put it in perspective, it comes to roughly 5–7% of the U.S. defense budget. Is 5% too much to spend on nuclear weapons? Several recent U.S. secretaries of defense have claimed that nuclear deterrence is the most important mission of the U.S. Department of Defense. So, put another way, is 5% too much to spend on the organization's highest priority mission? Reasonable people can disagree, but to many this seems to be a good value. Indeed, nuclear weapons have traditionally been prized because they are more economical than conventional weapons.[47] As Obama's Secretary of Defense Ash Carter pithily put it, "nuclear weapons don't actually cost that much."[48]

Other critics of U.S. modernization plans charge that building a new generation of nuclear weapons will fuel an arms race with Russia and China.[49] But this is incorrect. Russia and China are already modernizing their forces, and this is occurring irrespective of decisions made in Washington. Indeed, Russia is finishing a modernization cycle and is building new, exotic nuclear weapons, such as a nuclear submarine drone, that have never been contemplated in Washington. When Secretary Carter was asked if U.S. nuclear modernization was stimulating an arms race, he replied, "we know they aren't having that effect, because the evidence is to the contrary."[50] Moreover, U.S. modernization may actually dissuade arms races, as potential rivals realize they have little hope of achieving nuclear parity with or superiority over the technologically-sophisticated United States.

Finally, some claim that U.S. nuclear modernization plans are inconsistent with its commitments under the NPT to work toward eventual nuclear disarmament. According to this argument, U.S. modernization will cause nuclear proliferation, but this is also incorrect.[51] President Obama took good faith steps toward reducing reliance on nuclear weapons, but, as he did so, Russia, China, and North Korea went in the opposite direction, expanding and modernizing their arsenals. Disarmament advocates often harangue Washington, London, and Paris because they know these democracies will listen to their complaints, but they are barking up the wrong tree. If they want to make progress on disarmament, the real obstacle is Moscow, Beijing, and Pyongyang. If other nations up nuclear weapons, Washington would almost certainly join them. So long as these other nations have nuclear weapons, however, the United States needs to possess a robust nuclear deterrent. Moreover, as argued above, doing so reinforces global nonproliferation, by giving U.S. allies the incentive to forego independent nuclear arsenals. Meanwhile, the United States continues to make progress toward nuclear disarmament in other ways with new initiatives to "create the conditions" for future disarmament.[52]

In sum, U.S. nuclear modernization is necessary for the continued effectiveness of the U.S. nuclear deterrent. Moreover, such measures are affordable and consistent with America's longstanding disarmament goals.

The threat posed by Russian, Chinese, and North Korean nuclear modernization

Not all modernization programs are created equal. The effect of a state's nuclear modernization on international security depends upon that state's objectives and strategy, in accordance with its broader geopolitical disposition. Russia, China, and North Korea are revisionist states dissatisfied with the prevailing status quo. Their broader strategies seek to disrupt or displace the U.S.-led order, and thus their defense capabilities, including nuclear weapons, are oriented toward these destabilizing objectives. Nuclear modernization programs in Moscow, Beijing, and Pyongyang, therefore, tend to undermine international security and stability.

Russia

Russia seeks to disrupt the rules-based international order.[53] Under President Putin, Russia has sought to reassert itself as a global great power and to ensure that any major international security issue cannot be decided without Russia at the table. In addition, it seeks to carve out a sphere of influence in Eastern Europe at the expense of the security and sovereignty of its neighbors. It views the spread of the U.S. alliance system and of democratic forms of government as a fundamental threat to the existence of its regime.[54] Consequently, it seeks to exploit Western vulnerabilities in order to divide democratic nations within themselves and against each other. To achieve these objectives, Russia employs a "hybrid" or "new-generation warfare" military strategy that combines activities across the full spectrum of conflict, from information operations at the low end to nuclear coercion at the high end.[55]

The nuclear component of this broader grand strategy is designed to deter NATO from intervening in Russia's sphere of influence, to coerce neighboring states, and to divide NATO allies against each other.[56] In recent years, Russia has increased reliance on nuclear weapons in its strategy.[57] Russian leaders have made explicit nuclear threats and have postured and exercised nuclear forces in ways we have not seen since the Cold War. During the Russian invasion of Ukraine, for example, Russian President Vladimir Putin said that "Russia is one of the leading nuclear powers," and that "it is best not to mess with us."[58] He also put nuclear weapons on alert and moved them into Kaliningrad and Crimea. The message was clear: if the West interferes with Russia's attempts to use force against its neighbors, the result could be nuclear catastrophe.

The most dangerous component of Russian nuclear strategy is the so-called "escalate-to-de-escalate" doctrine.[59] This is simply the idea, widely acknowledged by Russian experts, that Russia will rely on nuclear weapons, including limited nuclear strikes, to offset the conventional superiority of the United States and NATO. This strategy sounds reasonable enough if it were only a defensive strategy to defend the Russian homeland from invasion. The concern is that, like in Ukraine, it will be used as a nuclear backstop to bolster future conventional aggression. Russia could invade a NATO member and then use threats of early nuclear escalation to deter a unified NATO response. Short of invasion, it can continue to use the threat for quotidian nuclear coercion of NATO members over foreign policy decisions of which it disapproves, such as regional states hosting NATO military capabilities.

To support this strategy, Russia is modernizing its nuclear arsenal. It is currently finishing a modernization cycle of its strategic nuclear forces. The Kremlin is building a new generation of non-strategic nuclear weapons, expanding its asymmetric advantages over NATO in this space, and giving credibility to its "escalate-to-de-escalate" approach.[60] Finally, it is developing exotic nuclear weapons, including nuclear-armed hypersonic glide vehicles, nuclear-powered nuclear cruise missiles, and nuclear-armed submarine drones.[61] The latter is designed as a system of pure terror to hold at risk Western port cities with the threat of a stealthy, large-yield, and highly radioactive strategic attack. These weapons contribute to Russia's strategic goals of weakening and terrorizing the free world and, as such, they undermine global security and stability.

China

While Russia is the more dangerous near-term threat, China is the greater threat to the global order over the long term. Russia can only disrupt the rules-based order, but China may seek to displace it. China endeavors to gradually replace the United States as the global hegemon by 2049.[62] The effort begins regionally in Asia where Beijing seeks to expel the United States as a regional power and establish itself as the region's hegemon. Globally, China is expanding its political and economic influence in every region of the world through its Belt and Road Initiative (BRI) and Made in China 2025 plans.[63] Its state-led capitalist approach seeks to prey on the rules-based economic order at the expense of open market democracies. It is also consciously or unconsciously exporting this state-led capitalist model and seeking to make the world safe for autocracy. Therefore, the rise of China is a profound threat to the rules-based order.

China supports this strategy with an ambitious military modernization program. In East Asia, its anti-access/area denial (A2/AD) approach seeks to undermine U.S. power projection capabilities and enable China to pursue territorial claims in the East and South China Seas, while providing coercive leverage vis-à-vis other regional actors.[64] In recent years, the balance of power in Asia has shifted and many question the ability of the United States to defend longstanding allies.[65] China is also expanding its global military footprint with overseas bases and military exercises in other world regions.

China is also building nuclear forces designed to contribute to these revisionist aims. While China claims to desire only a "lean and effective" deterrent, it has gradually expanded and modernized its forces in recent years.[66] Beijing is clearly attempting to develop a greater ability to hold the U.S. homeland at risk, to increase U.S. vulnerability to a large-scale nuclear exchange, and to deter U.S. military action, including in defense of its allies. In the past decade alone, the number of Chinese strategic nuclear weapons that could reach the continental United States has more than doubled. And U.S. intelligence has recently estimated publicly that China's nuclear arsenal will at least double again in the coming decade.[67] In addition to its strategic forces, China possesses a large stockpile of medium- and intermediate-range missiles that are capable of delivering nuclear weapons.[68] These capabilities contribute to China's A2/AD strategy and could be used to blunt U.S. and allied military forces in their attempts to defend against possible Chinese aggression. While China espouses a formal nuclear No First Use Policy, Chinese experts acknowledge that there is a narrow range of contingencies, such as a major war with the United States in East Asia, in which China might very well use nuclear weapons first. In the future, China may leverage its substantial resources to sprint to nuclear parity, potentially igniting an arms race with the United States and undermining U.S. extended deterrence and assurance in Asia.

North Korea

Relative to the Russian and Chinese great powers above, the rogue North Korean regime is not as threatening to the very fabric of the current international order.

Still, North Korea's nuclear program is a challenge to the rules-based order. Pyongyang has flagrantly violated the nonproliferation regime and its weapons threaten the security of the free world.

Scholars debate North Korean goals. Some argue that North Korea only wants to defend its brutal totalitarian regime from external intervention; others maintain that Pyongyang is pursing the more aggressive and revisionist goal of eventually reuniting the Korean peninsula under its own rule.[69] Regardless, North Korean behavior threatens almost every aspect of the rules-based order. It has repeatedly engaged in armed aggression against its neighbors from initiating the Korean War in the 1950s to sinking a South Korean warship in the 2010s. It cheats on global economic rules engaging in currency counterfeiting and black market smuggling activities. It also conducts massive human rights abuses at home.

North Korea's military strategy seeks to retain the capability to deter external invasion and, if necessary, to fight a full-scale war on the Korean Peninsula. North Korean nuclear capabilities contribute to these goals. It is publicly estimated that North Korea has the ability to deliver close-range, short-range, and medium-range missiles to U.S. allies and bases in Asia, while possessing enough fissile material to build between 30 and 60 nuclear warheads.[70] In addition, it is on the verge of developing an ICBM capability to hold the U.S. homeland at risk with the threat of nuclear war.[71]

While some hope that the international community may be able to negotiate the denuclearization of North Korea, it seems more likely that North Korea will remain a nuclear power for the foreseeable future. If North Korea's nuclear modernization continues, it will possess a more reliable means of threatening nuclear war against the United States and its allies at the core of the rules-based order. North Korean nuclear weapons may also act as a shield to external intervention, deterring the United States and incentivizing Pyongyang's aggression against South Korea or Japan. North Korea's nuclear weapons may also weaken U.S. extended nuclear deterrence and assurance, possibly persuading Tokyo and Seoul to reconsider their nuclear options. More broadly, North Korean possession of nuclear weapons undermines the NPT. North Korea may decide to once again export dangerous nuclear or missile technology. Pyongyang's nuclear possession may create a dangerous precedent as other nonnuclear nations hope to follow the same path of cheating on the NPT, building nuclear weapons, and ultimately being welcomed as de facto members of the nuclear club.

In sum, Russia, China, and North Korea are autocratic, revisionist challengers to the rules-based international order. As such, their nuclear programs threaten core pillars of the system of global stability and security that has reigned for the past several decades. In short, Russian, Chinese, and North Korean nuclear modernization programs undermine global stability and security.

The role of other nations' nuclear forces

To this point, we have reviewed the effect of the modernization programs of the United States and its adversaries, but what about the nuclear programs of other nations? This list of nuclear-armed powers includes America's nuclear-armed allies, Britain and France, and the three nuclear-armed states outside of the NPT framework: Israel, India, and Pakistan.

Like the United States, Britain and France have been status quo actors at the core of the Western liberal order since the end of World War II. Therefore, British and French nuclear weapons have, on balance, done more to reinforce, rather than undermine, sources of global stability. These allied nuclear forces have bolstered the capabilities of the Western alliance, thereby increasing its credibility and demonstrating alliance burden-sharing. Britain and France can deter direct attacks on their homeland. Nuclear weapons in these two nations also complicate Russia's nuclear targeting and make it possible that Russia will suffer a nuclear attack even if it succeeds in de-coupling America's nuclear weapons from the security of Europe.

However, British and French nuclear weapons have not had a comparable effect on global security. Due to America's global power projection capabilities and its vast international commitments, U.S. nuclear weapons have been more influential in underwriting European and global stability. As Albert Wohlstetter argued decades ago, at the end of the day, the nuclear balance in Europe rests primarily on U.S. and Russian nuclear arsenals.[72]

Moreover, while the United States was working hard to stop nuclear proliferation, France had a troubling history of using its nuclear capabilities to fuel horizontal proliferation. During the Cold War, France provided sensitive nuclear assistance to nonnuclear states, such as Israel and Pakistan.[73] Fortunately, France's nonproliferation policies have greatly improved since that time.

Britain and France are also undergoing their own nuclear modernization efforts and these capabilities will continue to make positive contributions to nuclear deterrence in Europe. It is ultimately the United States, however, which will continue to bear the greatest responsibility for using its nuclear weapons to deter great power war and halt the spread of the world's deadliest weapons.

India, Pakistan, and Israel are three states that never signed the NPT and then went on to build nuclear weapons. Unlike the other nuclear-armed states reviewed above, these states sit at the periphery of the U.S.-led global order. None of these states are formal U.S. allies, but they are not sworn enemies either. Washington, therefore, does not have a formal obligation to extend nuclear deterrence to these states and these states are not mentioned in the U.S. NPR as possible targets for U.S. nuclear weapons.

Israel is a democracy and a close, although not formal, American security partner. The United States failed in its bid to prevent Israel from developing nuclear weapons in the 1960s. Israel was, however, the repeated victim of conventional aggression by its conventionally superior and autocratic Arab neighbors in the past and its nuclear program has deterred subsequent invasions and generally contributed to Middle Eastern stability. Unlike Iran's advancing nuclear program today, Israel's nuclear weapons did not spur other regional states to develop their own nuclear weapons in response.

Nuclear programs in India and Pakistan are also contrary to the global nonproliferation order due to their acquisition outside of the NPT framework and because of a series of subsequent, dangerous nuclear crises in South Asia. The United States has maintained good relations with both India and Pakistan, but the orientation of these states may be shifting in this new era of great power competition. The Chinese-Pakistani strategic partnership seems to be deepening. And India is cooperating more closely with the United States and its democratic allies in Asia as a counterbalance to China's growing power. If these trends continue, India's nuclear forces may increasingly serve on balance to reinforce the sources of stability in the prevailing global order. While we can hope that Pakistan does not use its nuclear forces to complement China's revisionist ambitions.

Still, the existence of three nuclear powers outside of the NPT framework is somewhat problematic for efforts to develop a global, rules-based nonproliferation regime and there have been discussions in Washington about how best to mainstream these programs into the broader rules-based nuclear order.

Conclusion

Does nuclear modernization promote international security? This chapter argues that it depends. Previous scholarship has focused too narrowly on whether the technology itself is inherently stabilizing or destabilizing, but we argue that we can only make sense of the technology when it is situated within its proper geopolitical context. Different actors with different goals use the same technology in different ways. The status-quo-oriented United States has sought to create, advance, and defend the rules-based international order that has governed the global system since 1945. Thus, its nuclear modernization program tends to promote this overall system of international stability and security.

Several autocratic revisionist actors, on the other hand, are threatened by this rules-based system and believe that it does not reflect their interests and priorities. Russia, China, and North Korea are autocracies that are dissatisfied with significant aspects of the reigning order. Accordingly, they are pursuing strategies to revise the international status quo. Their military postures are oriented toward disrupting the current global order, and it is therefore likely that their nuclear modernization programs will diminish international security and stability.

This argument has scholarly and policy implications. Scholars should move beyond abstract discussions about whether nuclear weapons (or modernization programs) are stabilizing or not and focus more on the important variation caused by the strategies of weapons possessors. States use nuclear weapons in different ways and this has implications for how nuclear weapons affect international politics. Current political science models seek to understand the likely effects of nuclear acquisition in Iran, for example, by conducting quantitative analysis of all instances of past proliferation.[74] But aggregated data that contains information about U.S., British, and French behavior with nuclear weapons will likely be misleading in attempting to understand how a nuclear Iran will act. Like the revisionist autocracies reviewed above, Iran is another power with a foreign policy objective of resisting the United States and the prevailing international order. It is likely, therefore, that China or North Korea are more accurate models for imagining the behavior of a future nuclear-armed Iran.

In the realm of policy implications, we conclude that the United States must follow through on its modernization plans. U.S. nuclear weapons play a unique role in protecting the free world and preventing the spread of nuclear weapons. The vast majority of states in the international system have a strong interest in the continued salutary effects of a robust U.S. nuclear deterrent.

On the other hand, Russian, Chinese, and North Korean nuclear modernization poses a threat to international security and stability. The United States and its allies must work to maintain a favorable balance of power over these autocratic challengers. They should seek to retard Russian and Chinese nuclear modernization through arms control agreements, export controls on advanced technology, and other mechanisms. In North Korea, the international community must continue to work toward complete and verifiable denuclearization. In the meantime, the United States and its allies must put in place a robust nuclear strategy and posture to deter possible aggression from Beijing, Moscow, and Pyongyang.

In sum, U.S. nuclear weapons have served as the unique guarantor of international security since 1945. Let us hope, therefore, that U.S. current nuclear modernization plans are sufficient to continue to preserve the peace in the decades to come.

Notes

1 Kenneth N. Waltz, *The Spread of Nuclear Weapons: More May Be Better*, Adelphi papers, no. 171 (London: International Institute for Strategic Studies, 1981); Kenneth N. Waltz, "Nuclear Myths and Political Realities," *The American Political Science Review* 84, no. 3 (September 1990), 731–745; John J. Mearsheimer, "Why We Will Soon Miss the Cold War," *The Atlantic Monthly* 266, no. 2 (August 1990), 35–50; John J. Mearsheimer, "The Case for a Ukrainian Nuclear Deterrent," *Foreign Affairs* 72, no. 3 (Summer 1993), 50–66; Keir A. Lieber and Daryl G. Press, "Obama's Nuclear Upgrade: The Case for Modernizing America's Nukes," *Foreign Affairs* (July 6, 2011), www.foreignaffairs.com/articles/2011-07-06/obamas-nuclear-upgrade; Peter Huessy, "Why America Must Modernize Its Nuclear Forces," *War on the Rocks*, March 15, 2018, https://warontherocks.com/2018/03/why-america-must-modernize-its-nuclear-forces/; Elbridge Colby, "If You Want Peace, Prepare for Nuclear War: A Strategy for the New Great-Power Rivalry," *Foreign Affairs* 97, no. 6 (November–December 2018).

2 Katrina vanden Heuvel, "The *Nuclear Posture Review* Signals a New Arms Race," *The Washington Post*, February 13, 2018, www.washingtonpost.com/opinions/the-nuclear-posture-review-signals-a-new-arms-race/2018/02/13/de155e64-1018-11e8-8ea1-c1d91fcec3fe_story.html?utm_term=.d42c89c70585; Eric Schlosser, "The Growing Dangers of the New Nuclear-Arms Race," *The New Yorker*, May 24, 2018, www.newyorker.com/news/news-desk/the-growing-dangers-of-the-new-nuclear-arms-race.

3 G. John Ikenberry, *Liberal Leviathan: The Origins, Crisis, and Transformation of the American World Order* (Princeton, NJ: Princeton University Press, 2012); Matthew Kroenig and Ash Jain, *Present at the Recreation: A Global Strategy for Revitalizing, Adapting, and Defending the Rules-Based International Order*, Atlantic Council Strategy Papers (Washington, DC: Atlantic Council, 2019).

4 Matthew Kroenig, *The Logic of American Nuclear Strategy: Why Strategic Superiority Matters* (New York: Oxford University Press, 2018).

5 Ikenberry, *Liberal Leviathan*; Kroenig and Jain, *Present at the Recreation*.

6 Max Roser, "The Short History of Global Living Conditions and Why It Matters That We Know It," *Our World in Data*, https://ourworldindata.org/a-history-of-global-living-conditions-in-5-charts#why-it-matters-that-we-do-not-know-how-our-world-is-changing.
7 Freedom House, *Freedom in the World 2019: Democracy in Retreat* (Washington, DC: Freedom House, 2019), 1.
8 Waltz, *The Spread of Nuclear Weapons*; Robert Jervis, *The Illogic of American Nuclear Strategy*, Cornell Studies in Security Affairs (Ithaca, NY: Cornell University Press, 1984); Robert Jervis, *The Meaning of the Nuclear Revolution: Statecraft and the Prospect of Armageddon*, Cornell Studies in Security Affairs (Ithaca, NY: Cornell University Press, 1989); Mearsheimer, "Why We Will Soon Miss the Cold War," 35–50; Waltz, "Nuclear Myths and Political Realities," 731–745; Thomas C. Schelling, *Arms and Influence: With a New Preface and Afterword* (New Haven, CT: Yale University Press, 2008); Kroenig, *The Logic of American Nuclear Strategy*.
9 Kroenig, *The Logic of American Nuclear Strategy*, 4.
10 John Lewis Gaddis, *The United States and the Origins of the Cold War, 1941–1947* (New York: Columbia University Press, 1972); John Lewis Gaddis, *We Now Know: Rethinking Cold War History*, 1st edition (New York: Oxford University Press, 1997).
11 Gaddis, *The United States and the Origins of the Cold War, 1941–1947*; Gaddis, *We Now Know*; Mark Kramer, "Stalin, Soviet Policy, and the Establishment of a Communist Bloc in Eastern Europe, 1941–1949," in Mark Kramer and Vit Smetana (eds.), *Imposing, Maintaining, and Tearing Open the Iron Curtain: The Cold War and East-Central Europe, 1945–1989* (Lanham, MD: Lexington Books, 2014), 3–38.
12 Lawrence Freedman, "The First Two Generations of Nuclear Strategists," in Peter Paret (ed.), *Makers of Modern Strategy: From Machiavelli to the Nuclear Age* (Princeton, NJ: Princeton University Press, 1986), 735–745.
13 Ibid., 735–778.
14 Matthew Kroenig, *The Return of Great Power Rivalry: Democracy Versus Autocracy from the Ancient World to the United States, Russia, and China Today* (New York: Oxford University Press, 2020).
15 "Putin Deplores Collapse of USSR," *BBC News*, April 25, 2005, http://news.bbc.co.uk/2/hi/4480745.stm; Vladimir Socor, "Putin's Crimea Speech: A Manifesto of Greater-Russia Irredentism," *Eurasia Daily Monitor* 11, no. 56 (2014).
16 Paul Sonne et al., "'No First Use' Nuclear Policy Proposal Assailed by U.S. Cabinet Officials, Allies," *The Wall Street Journal*, August 12, 2016, www.wsj.com/articles/no-first-use-nuclear-policyproposal-assailed-by-u-s-cabinet-officials-allies-1471042014; David E. Sanger and William J. Broad, "Obama Unlikely to Vow No First Use of Nuclear Weapons," *The New York Times*, September 5, 2016, www.nytimes.com/2016/09/06/science/obama-unlikely-to-vow-no-first-use-of-nuclear-weapons.html.
17 United States Department of Defense, *Nuclear Posture Review* (Washington, DC: Department of Defense, February 2018), 52–55.
18 Matthew Kroenig, *Exporting the Bomb: Technology Transfer and the Spread of Nuclear Weapons*, Cornell Studies in Security Affairs (Ithaca, NY: Cornell University Press, 2010); Matthew Kroenig, "Force or Friendship? Explaining Great Power Nonproliferation Policy," *Security Studies* 23, no. 1 (2014).
19 Rebecca Davis Gibbons, "American Hegemony and the Politics of the Nuclear Nonproliferation Regime," PhD dissertation, A dissertation submitted to the Faculty of the Graduate School of Arts and Sciences of Georgetown University in partial fulfillment of the requirements for the degree of Doctor of Philosophy in Government, Georgetown University, Washington, DC, April 2016; Andrew J. Coe and Jane Vaynman, "Collusion and the Nuclear Nonproliferation Regime," *The Journal of Politics* 77, no. 4 (October 2015).
20 Waltz, *The Spread of Nuclear Weapons*.

21 Matthew Kroenig, "The History of Proliferation Optimism: Does It Have a Future?" *The Journal of Strategic Studies* 38, nos. 1–2 (2015).
22 Dan Reiter, "Security Commitments and Nuclear Proliferation," *Foreign Policy Analysis* 10 (2014), 61–80.
23 Philipp C. Bleek and Eric B. Lorber, "Security Guarantees and Allied Nuclear Proliferation," *Journal of Conflict Resolution* 58, no. 3 (2014), 438.
24 Clark A. Murdock and Jessica M. Yeats, *Exploring the Nuclear Posture Implications of Extended Deterrence and Assurance: Workshop Proceedings and Key Takeaways* (Washington, DC: Center for Strategic & International Studies, November 2009).
25 Reiter, "Security Commitments and Nuclear Proliferation," 65–73.
26 Matthew Karnitschnig, "German Bomb Debate Goes Nuclear," *Politico*, August 3, 2018, www.politico.eu/article/german-bomb-debate-goes-nuclear-nato-donald-trump-defense-spending/.
27 Michelle Ye Hee Lee, "More than Ever, South Koreans Want Their Own Nuclear Weapons," *The Washington Post*, September 13, 2017, www.washingtonpost.com/news/worldviews/wp/2017/09/13/most-south-koreans-dont-think-the-north-will-start-a-war-but-they-still-want-their-own-nuclear-weapons/?utm_term=.affa7f250c20.
28 Clark Murdock and Thomas Karako, *Thinking About the Unthinkable in a Highly Proliferated World* (Washington, DC: Center for Strategic & International Studies, July 2016).
29 Kroenig, *Exporting the Bomb*.
30 Ibid.
31 Matthew Kroenig, "U.S. Nuclear Weapons and Non-Proliferation: Is There a Link?" *Journal of Peace Research* 53, no. 2 (2016), 166–179.
32 Kroenig, *The Logic of American Nuclear Strategy*, 159–177.
33 United States Department of Defense, *Nuclear Posture Review*, 45.
34 Ibid., 46.
35 Ibid., 47.
36 Ibid., 56–58.
37 Ibid., 61.
38 Matthew Kroenig, "How to Approach Nuclear Modernization? A U.S. Response," *Bulletin of the Atomic Scientists* 71, no. 3 (2015), 16–18.
39 United States Department of Defense, *Nuclear Posture Review*, 49.
40 Ibid., 49–50.
41 Ibid., 50.
42 Ibid., 56–64.
43 Anna Péczeli, "Continuity and Change in the Trump Administration's Nuclear Posture Review," *Bulletin of the Atomic Scientists*, February 20, 2018, https://thebulletin.org/commentary/continuity-and-change-in-the-trump-administrations-nuclear-posture-review/; United States Department of Defense, *Nuclear Posture Review*.
44 United States Department of Defense, *Nuclear Posture Review*.
45 John R. Harvey et al., "Continuity and Change in U.S. Nuclear Policy," *RealClearDefense*, February 7, 2018, www.realcleardefense.com/articles/2018/02/07/continuity_and_change_in_us_nuclear_policy_113025.html.
46 Kroenig, *The Logic of American Nuclear Strategy*, 178–187.
47 Ibid.
48 Ashton B. Carter, "Remarks by Deputy Secretary of Defense Carter at the Aspen Security Forum at Aspen, Colorado," *U.S. Department of Defense*, July 18, 2013, https://archive.defense.gov/Transcripts/Transcript.aspx?TranscriptID=5277.
49 Kroenig, *The Logic of American Nuclear Strategy*, 143–158.
50 Ashton B. Carter, "Remarks on 'Sustaining Nuclear Deterrence'," *U.S. Department of Defense*, September 26, 2019, https://dod.defense.gov/News/Speeches/Speech-View/Article/956630/remarks-on-sustaining-nuclear-deterrence/.

51 Matthew Kroenig, "Think Again: American Nuclear Disarmament," *Foreign Policy*, September 3, 2013, https://foreignpolicy.com/2013/09/03/think-again-american-nuclear-disarmament/.

52 Christopher A. Ford, "Creating the Conditions for Nuclear Disarmament: A New Approach," U.S. Department of State, March 17, 2018, www.state.gov/remarks-and-releases-bureau-of-international-security-and-nonproliferation/creating-the-conditions-for-nuclear-disarmament-a-new-approach/.

53 Angela Stent, *Putin's World: Russia Against the West and with the Rest* (New York: Hachette Book Group, 2019); Matthew Kroenig, "Facing Reality: Getting NATO Ready for a New Cold War," *Survival* 57, no. 1 (2015), 49–70.

54 Thomas E. Graham, "The Sources of Russian Conduct," *The National Interest*, August 24, 2016, https://nationalinterest.org/feature/the-sources-russian-conduct-17462.

55 Phillip Karber and Joshua Thibeault, "Russia's New-Generation Warfare," *Army Magazine* 66, no. 6 (June 2016), 60–64; Valery Gerasimov, "The Value of Science Is in the Foresight: New Challenges Demand Rethinking the Forms and Methods of Carrying out Combat Operations," trans. Robert Coalson, *Military Review: The Professional Journal of the U.S. Army* (January–February 2016), 23–29.

56 Keith B. Payne, "The Emerging Nuclear Environment: Two Challenges Ahead," *RealClearDefense*, January 3, 2019, www.realcleardefense.com/articles/2019/01/03/the_emerging_nuclear_environment_two_challenges_ahead_114074.html.

57 Matthew Kroenig, *The Renewed Russian Nuclear Threat and NATO Nuclear Deterrence Posture* (Washington, DC: Atlantic Council, February 2016).

58 Paul Sonne, "As Tensions with West Rise, Russia Increasingly Rattles Nuclear Saber," *The Wall Street Journal*, April 5, 2015, www.wsj.com/articles/as-tensions-with-west-rise-russia-increasingly-rattles-nuclear-saber-1428249620.

59 Matthew Kroenig, *A Strategy for Deterring Russian Nuclear De-Escalation Strikes* (Washington, DC: Atlantic Council, April 2018), 14.

60 Hans M. Kristensen and Matt Korda, "Russian Nuclear Forces, 2019," *Bulletin of the Atomic Scientists* 75, no. 2 (2019), 73–84; Kroenig, *A Strategy for Deterring Russian Nuclear De-Escalation Strikes*, 7; United States Department of Defense, *Nuclear Posture Review*, 9.

61 Kristensen and Korda, "Russian Nuclear Forces, 2019"; Kroenig, *A Strategy for Deterring Russian Nuclear De-Escalation Strikes*, 7; United States Department of Defense, *Nuclear Posture Review*, 9.

62 Graham Allison, "What Xi Jinping Wants," *The Atlantic*, May 31, 2017, www.theatlantic.com/international/archive/2017/05/what-china-wants/528561/; United States Department of Defense, *Summary of the 2018 National Defense Strategy of the United States of America: Sharpening the American Military's Competitive Edge* (Washington, DC: Department of Defense, 2018), 2; Hal Brands, "China's Master Plan: A Global Military Threat," *Bloomberg*, June 10, 2018, www.bloomberg.com/opinion/articles/2018-06-10/china-s-master-plan-a-global-military-threat.

63 Andrew Chatzky and James McBride, "China's Massive Belt and Road Initiative," *Council on Foreign Relations*, last updated May 21, 2019, www.cfr.org/backgrounder/chinas-massive-belt-and-road-initiative; Gal Luft, *Silk Road 2.0: U.S. Strategy Toward China's Belt and Road Initiative*, Atlantic Council Strategy Papers (Washington, DC: Atlantic Council, 2017).

64 Evan Braden Montgomery, "Contested Primacy in the Western Pacific: China's Rise and the Future of U.S. Power Projection," *International Security* 38, no. 4 (Spring 2014), 115–149.

65 National Defense Strategy Commission, *Providing for the Common Defense: The Assessment and Recommendations of the National Defense Strategy Commission* (Washington, DC: United States Institute of Peace, November 2018).

66 Fiona S. Cunningham and M. Taylor Fravel, "Assuring Assured Retaliation: China's Nuclear Posture and U.S.-China Strategic Stability," *International Security* 40, no. 2 (Fall 2015), 7–50.
67 Robert P. Ashley, "Russian and Chinese Nuclear Modernization Trends: Remarks at the Hudson Institute," *Defense Intelligence Agency*, May 29, 2019, www.dia.mil/News/Speeches-and-Testimonies/Article-View/Article/1859890/russian-and-chinese-nuclear-modernization-trends/.
68 Hans M. Kristensen and Robert S. Norris, "Chinese Nuclear Forces, 2018," *Bulletin of the Atomic Scientists* 74, no. 4 (2018), 289–295.
69 Bruce Klingner, "U.S. Should Counter North Korea's Strategic Objectives," *The Journal of East Asian Affairs* 32, no. 1 (Spring–Summer 2018), 4.
70 Hans M. Kristensen and Robert S. Norris, "North Korean Nuclear Capabilities, 2018," *Bulletin of the Atomic Scientists* 74, no. 1 (2018); Klingner, "U.S. Should Counter North Korea's Strategic Objectives," 3–4.
71 Klingner, "U.S. Should Counter North Korea's Strategic Objectives," 4.
72 Albert Wohlstetter, "Nuclear Sharing: NATO and the N+1 Country," *Foreign Affairs* 39, no. 3 (April 1961), 355–387.
73 Kroenig, *Exporting the Bomb.*
74 Erik Gartzke and Matthew Kroenig, "Nukes with Numbers: Empirical Research on the Consequences of Nuclear Weapons for International Conflict," *Annual Review of Political Science* 19 (2016), 397–412.

11 Afterword

Philip M. Baxter and Aiden Warren

The following attempts to outline the nuclear modernization efforts persisting in the United States, Russia, and China, and to discuss the strategic and policy implications of these developments. As conveyed earlier, this volume has sought to provide not only a technical context, but also to frame the likely effects nuclear modernization could have on the relations between these nuclear weapon powers and the larger impact upon efforts to curb nuclear weapons, both in terms of horizontal and vertical proliferation. The chapters have been arranged so as to inform a variety of stakeholders, from academics to policy-makers, by connecting analytical and normative insights from developments within the nuclear weapons arena. Such insights have resulted in discreet observations and tangible recommendations outlined by the individual authors who have contributed to this edited collection.

This final chapter will not attempt to summarize the observations, findings, and recommendations made by the authors. Rather, we will endeavor to be forward-looking, examining potential scenarios and possible futures that could arise within three distinct domains; given the trajectory of nuclear weapons modernization and outlook discussed over the previous ten chapters. These three domains include: strategic stability, great power conflict, and arms control and nonproliferation.

Strategic stability

Strategic stability is an oft used term plagued by a lack of consensus in its definition. Deterrence is at the core of strategic stability and is predicated on threat. The threat of denying an opponent their objectives through undertaking a course of action, or the threat of punishment in kind if they (the opponent) do undertake an action. Strategic stability, while less well defined, is functionally the existence of parity between states such that both feel relatively secure given its opponent's insecurity. Aside from this rather broad definition, it is also important to consider the components that make up strategic stability. Adam Stulberg and Lawrence Rubin outline two. Firstly, strategic stability is the "Condition in which adversaries understand that altering military force posture in response to vulnerability – whether to avoid being emasculated or to preempt one's opponent – would be either futile or foolish."[1] Secondly, "strategic stability reflects the ease which nuclear-armed adversaries can return to stable relations after a period of escalation."[2]

With this framework in place, the pertinent question, therefore, asks how might nuclear modernization programs directly impact strategic stability? One area in which modernization can impact, and to some extent has already done so, applies to when the existing strategic stability frameworks contribute to the collapse of arms control agreements which limit the number of strategic nuclear armaments or the types of armaments. In the broader geopolitical and security landscapes, it is evident that such dynamics are playing out in the present between the U.S. and Russia. As both countries view the other as pushing the bounds of treaty restrictions or introducing technologies that shift strategic parity, the strategic calculi are adjusted, and stability is weakened.

Of course, a transitioning international system introduces interesting dynamics for existing strategic stability frameworks. The existing framework evolved in a bipolar international system during the Cold War and centered on deterring large-scale conventional and nuclear conflict between the U.S. and Russia. The arms treaties signed near the end of the Cold War locked in this framework for the proceeding unipolar period of U.S. dominance. While the present power structure and its trajectory are extensively debated, it is plausible to postulate that a shift away from unipolarity will engender the rise of new and more complex strategic stability challenges. Clearly, if the United States seeks to maintain the present security framework and its primacy, the management of multiple emerging challengers to its influence will be required. At present, most strategists see China as the most worrisome potential challenger to U.S. hegemony, particularly given China's rapid rise and size. Similarly, India's trajectory over the next few decades is likely to increase. And while it has a less contentious relationship with the U.S., it has its own strategic stability concerns emanating from its relationship with Pakistan and China. In the context of Russia, it sees itself as a re-emerging powerful state who has sought to reassert itself and its sphere of influence over the last decade. In addition to continued vertical proliferation by other states not discussed in this volume, it is evident that nuclear modernization efforts and the emergence of new technologies will present further stress and complexities to the existing strategic paradigms – and exacerbated even further if they exist within and across contentious state relationships.

Indeed, emerging technologies and their integration with previous nuclear weapons missions introduce new challenges for existing frameworks of strategic stability. As mentioned in the above, acceptance of the fact that substantive force posture changes would be a futile endeavor is a key element of strategic stability. However, as states seek the integration of new military technologies to bolster their forces and expand their deterrent capabilities, the perspective that the emerging technology is distinct from nuclear forces (or other deterrents), and thus not affecting strategic stability is significantly tested. An example of this would be the development of hypersonic weapons, which given their capacity for a rapid, long-distance response with conventional armaments, can be viewed as a more usable weapon system than the nuclear arsenal.

As new technologies become available and the domains for potential conflict expand, the rethinking of strategic stability paradigms will need to incorporate not only new actors, such as states with lower nuclear weapon counts, but also states who could potentially upset existing stable relations through asymmetrical means. In this regard, a reassessment and rethinking into how new technologies will interact and impact threat perceptions when coupled is imperative. Adding additional nuclear weapons to large stockpiles is not as debilitating to strategic stability when compared to the integration of new classes of weapons that introduce capabilities. Here, an opponent is put in a perceived (and potentially severe) disadvantaged position which cannot be easily overcome by bolstering its own capabilities, even if more rudimentary. Technologies that could be introduced to the nuclear weapons programs of states over the next few decades could have this potential impact, and as such, the careful consideration that such new developments will pose on existing strategic outlooks and potential scenarios is critical.

Great power conflict

Great power conflict has once again emerged as a nascent issue for international security. The concern over the return to a period of great power conflict has sharply risen over the last few years. While some would point to a resurgent Russia and concerns regarding their renewed quest for stature and regional influence, the primary concern for many Western policymakers, strategists and analysts has been the rapid and monolithic rise of China. How nuclear weapon modernization efforts will impact these great power dynamics and the prospects for conflict will remain a pertinent question in the coming decades.

The twentieth century was characterized by two broad periods: a multipolar international system marked by great power conventional war, and a bipolar power structure characterized by arms racing and proxy conflicts. The prospect of a tri-polar system, with the United States, China, and Russia in competitive and relatively equal positions, or a multi-polar system with various spheres of competing influence introduces new and potentially dangerous scenarios.

As states seek out security, either through the promotion of their own forces in anticipation of future conflict or in response to an adversary's development of new technologies or military assets, arms racing and greater uncertainty in intentions is likely to build. One area in which scholarship and broader discourse has struggled in assessing great power competition pertains to the extent to which emerging technologies will impact power dynamics and perceptions of shifting power. While nuclear weapons have been fully integrated into this thinking of great power competition, the processes of modernization and new technology associated with their deployment are less understood.

Nuclear modernization in the context of the refurbishment of older weapons may not escalate and intensify such debates, but other dynamics discussed throughout this collection may. In particular, the advancement of technologies to expand the capabilities or mission set of nuclear forces would introduce greater uncertainty. Additionally, linkages between nuclear forces and other weapon platforms could also present unique escalatory challenges between great powers. For example, cyber-attacks against nuclear command and control centers or the further development, deployment, and usage of hypersonic glide weapons.

Many have argued that the U.S. and China are on a collision course of great power competition. The cause has been attributed to the contrasting political systems, perceived desire for greater power by China, China's desire to remake the international system to fit its needs, the desire of the U.S. to limit China's trajectory in the Indo-Pacific, a view of weakening vis-à-vis China within the U.S., the rapid economic advancement of China, and so forth. These perceptions of tensions are exacerbated by Chinese investment into and rapid advancement in bolstering its military capabilities, in particular, its maritime capacity and its information sectors.

We can imagine a scenario in which perceptions of an ongoing power transition motivates actors to take actions to either further the shifting power dynamics or to maintain the status quo. With investments by the U.S., China, and Russia (who sees itself as once again warranting a sphere of influence and in discussion pertaining to great power) into military assets and nuclear forces in particular, the potential for risk-taking grows as perceptions of an ongoing transitory period increase. Such risk-taking behavior could include meeting an opponent's moves with matching or using asymmetric capabilities as a means to maintain its status or to offset any gains by an opponent, with the latter being more likely to result in conflict.

The renewed discussions regarding great power competition provide a unique lens from which to assess modernization programs. In one regard, nuclear modernization can be thought of as merely being the process in which a state ensures its current arsenal is safe and well maintained. However, as this volume has discussed, current modernization efforts appear to be more focused on strengthening capabilities, introducing new agile tools in nuclear arsenals, and meeting perceived rising challenges. While one could look to diplomatic engagement, broader military activities, political posture, or other indicators for signs of fraying relations and greater consternation between great powers, the urgency and vast undertaking in modernizing nuclear forces provides evidence that states see great power competition as not a relic of the past, but rather a forthcoming period.

Arms control and nonproliferation

There has been a distinct shift in the arms control and nonproliferation agenda over the last three decades. During the Cold War, parallel tracks sought to both restrict the horizontal proliferation of nuclear weapons to new states as well as the bilateral approaches to limit vertical proliferation. In addition to limited weapon count, efforts were made to restrict delivery systems, classes of weapons, and other systems intended to advance deterrence strategies.

At the end of the Cold War, an effort was made to make dramatic cuts in the number of nuclear weapons the United States and the Soviet Union each possessed. Additionally, efforts were made to further stymie the development of new weapon systems, culminating in the Comprehensive Test Ban Treaty, which while not fully enforced, normalized the non-testing norm propagated by moratoriums on all nuclear testing by the Soviet Union in 1991 and the United States in 1993, stemming from the work of the Partial Test Ban Treaty.

Recent efforts in the nonproliferation regime have been fortifying against horizontal proliferation, through efforts such as the Proliferation Security Initiative, the Joint Comprehensive Plan of Action (JCPOA) to limit Iran's breakout capability, or to pressure North Korea to roll back its nuclear weapon development. Concurrently, the non-nuclear weapon states (NNWS) have grown increasingly frustrated with the lack of movement on disarmament by the nuclear weapon states (NWS), culminating with the Treaty on the Prohibition of Nuclear Weapons in 2017.

As has been discussed in this volume, modernization presents new challenges to the existing arms control and nonproliferation regimes. As states seek to develop new weapons, norms against testing will be challenged as states seek to ensure the viability of such weapons. Arms racing based on perceived vulnerability stemming from new nuclear capabilities will put pressure on the ability of nuclear weapon states to negotiate further arms reductions, as well as further ostracizing non-nuclear states who view the lack of movement on disarmament as weakening the arguments for continued adherence to the nonproliferation regime. As current nuclear weapon states rearm and introduce new weapon systems, non-nuclear weapon states will continue to question the commitments within the NPT by the recognized weapon states to disarm.

Given the gap between nuclear weapon states' interest in disarmament, coupled with the rapidly approaching development of technologies for the nuclear weapon enterprise, scenarios can be envisaged in which greater instability arises as states attempt to adjust their deterrence strategies to cope with perceived vulnerabilities. These technologies are not limited to new nuclear weapons or an increase in the number of nuclear weapons. Highly accurate targeting capabilities of nuclear weapons are argued to allow for smaller, and thus more useable weapons; and which seem to have perpetuated both a similar need, as well as non-nuclear weapons of similar capabilities. Technologies that allow for striking command and control capabilities through cyber-attacks coupled, potentially, with artificial intelligence could motivate preemptive action. However, while best practices can be established in cyber-space and defenses bolstered to the greatest extent possible, these types of capabilities are significantly limited in their capacity to regulate via arms controls.

Another scenario that could potentially plague nuclear disarmament and nonproliferation in the near future is the aging of previous stewards of disarmament negotiations, in particular between the United States and Russia. As a new generation steps in to take up the mantle of disarmament, long-standing relationships and previous engagement can help to smooth negotiations. Given this, it is plausible that new negotiations will be sought to discuss emerging technologies' integration into nuclear capabilities. Of course, the duration of these efforts and hurdles faced will likely be greater given the diminished experience lost in the context of this class of negotiations; and by this we mean the extent to which new ideas and programs will need to be established rather than deciding on a number nuclear weapons or delivery systems to deploy.

A final issue that has arisen over the last decade is the democratization and accessibility of information. In the past, negotiations and the critical element of enforcing the resulting treaties have been national technical means, such as satellite imagery monitoring. With data readily available and quickly accessed, a new group of stakeholders has entered the equation of arms control and disarmament activities. Non-state actors such as think tanks or individual subject matter experts can release details of a state's nuclear weapons program that has not been the purview of those outside of government and usually compartmentalized to a very small portion of individuals in a government. As states seek to conduct negotiations, greater transparency on the part of the state parties may be necessary as a larger and more public element, but may even release details that hurt negotiating positions. This may result in negotiations being slower to start as states focus on securing their positions and developments prior to negotiations, rather than conducting the actions in tandem.

Closing thoughts

Nuclear modernization efforts currently planned or already underway present new and challenging scenarios for international security. Rather than a trajectory towards a perceived reduction in the utility of nuclear weapons, the opposite appears to be true. Value beyond simply deterring the use of nuclear weapons by other states appears to no longer be the case. Upgrading systems, introducing new capabilities, the intermingling of new technologies, and advancing new strategic models are all indicative of the elevated importance (and reliance) of nuclear weapons. The next decade will see a more contentious international environment, particularly between great power states, than the previous three decades and will be marked by increasing investment in nuclear enterprises.

Given this pessimistic outlook, what can be done? If we start from the premise that arms control and limitation agreements are unlikely in the near-term as states seek to modernize their nuclear forces, how do we limit potential escalatory situations?

One area where states undergoing modernization can mitigate arms racing and limit potential escalation is to advance transparency in the number of nuclear weapons they are refurbishing and deploying. While some bilateral agreements require reporting, new classes of weapons may not be viewed as requiring disclosure under existing frameworks or may be developed without testing and requiring disclosure. Ensuring the level of development of nuclear weapons will help states communicate the risks they perceive to be pertinent to stability.

A second, more difficult area of transparency would be in the capabilities being developed. Greater transparency and disclosure of capabilities, historically done through testing, allow for a state to plan accordingly and to not over-extend. During the Cold War, nuclear weapon testing and missile testing allowed for clear communications between the U.S. and the Soviet Union, and eventually China, to indicate the level of reliability and power obtained through its development program. While a resumption of nuclear weapon testing should not be undertaken, preforming tests of support and delivery systems in a way that would allow for assessment would tamper perceptions of gaps in capabilities and help to stabilize arms racing.

Lastly, and at this stage likely the most difficult, a renewed dialogue at the government-to-government level to help facilitate an understanding of motivations, developments, and perceptions is needed. Of course, given the challenges emanating from an apparent intensification in power politics, the task ahead and opportunity for such fora will no doubt be very difficult to attain, but certainly not impossible.

Notes

1 Lawrence Rubin and Adam Stulberg, *The End of Strategic Stability: Nuclear Weapons and the Challenge of Regional Rivalries* (Washington D.C.: Georgetown Press, 2018).
2 Ibid.

Index

Note: Page numbers in **bold** indicate a table on the corresponding page.

ABM *see* Anti-Ballistic Missile (ABM) Treaty
academic debate 61–66
Air Force 16–19, 94–95, 117–118, 152–153
air-launched cruise missiles (ALCMs) 15–17, 45–46, 54, 95, 115, 117; and international security 151, 161, 163, 178–179; *see also* long-range standoff missile (LRSO)
ALCM *see* air-launched cruise missiles (ALCMs)
Algeria 177
Anti-Ballistic Missile (ABM) Treaty 42–43, 52, 83n81, 84n82, 84n88
arms control 6–13, 32–33, 50–53, 90–91, 98–100; cold shoulder 159–160; and international security 146–147, 162–163, 191–192, 194–196; and NATO nuclear modernization 111–112, 120–121
arms race 1–2, 80n10, 99–101, 158–159, 179–180
arsenals 1–9, 12–16, 19–20, 26–29, 34–35; and Chinese nuclear strategy 63–64; excessive 147–149; and International Relations theory 125–126, 135–137; and international security 145–150, 154–160, 164–165, 172n92, 173–174, 177–182; and NATO nuclear modernization 115–117; and the nuclear nonproliferation regime 89–90; quantitative changes to 50–53; the role of the U.S. nuclear arsenal today 130–133; and U.S.-Russian bilateral disarmament 42–43, 46–48, 50–53
Asia 15–16, 19, 34, 40, 132, 149; and international security 174–176, 182–183, 185
Australia 88, 175
aviation, strategic *see* strategic aviation

ballistic missiles, intercontinental 1, 26–27, 29–31, 41, 64, 148, 178
ballistic missiles, land-based 7, 17–18
ballistic missiles, submarine-launched 27–29, 41, 77–79, 86, 110, 148, 178
ballistic missile submarines 18–19, 94–95, 115–117, 160–161; sea-launched 31
Boeing 18
Britain *see* United Kingdom

CEP *see* circular error probable (CEP) 26, 75, 83n76
CFE Treaty *see* Conventional Armed Forces in Europe Treaty (CFE Treaty)
changes 91–93, 124–125, 128–129; qualitative changes to arsenals 50–53
China 1–3, 5–9, 12, 52, 191–194, 197; and the academic debate 62–66; attitudes towards the bomb 133–137; and conditions of use 70–72; and evidence of transition 66–77, **67**; and force structure 74–77; and International Relations theory 125, 127, 130, 132, 140; and international security 145–149, 152, 155–164, 173–179, 183–186; and NATO nuclear modernization 119; and the nuclear nonproliferation regime 94, 96, 98, 100; nuclear strategy 61–79, 80n10, 81n25, 83n61, 83n75, 83n78, 84n82, 84n84, 84n88, 84n94, 85n98; and objectives 67–69; and targeting 72–74; threat posed by modernization of 180, 182; views on deterrence 135–137

Cold War 6–7, 12–13, 192, 194, 196–197; and International Relations theory 125–127, 136–137; and international security 156–158, 174–177; and NATO nuclear modernization 106–108, 116–117; and the nuclear nonproliferation regime 86–87, 90–92; rethinking Cold War nuclear deterrence 130–133; and Russia's nuclear modernization 26, 29–31, 33–34; and U.S.-Russian bilateral disarmament 47, 59
Comprehensive Nuclear Test Ban Treaty (CTBT) 77, 84n84, 159
conditions of use 67, **67**, 70–72
Conventional Armed Forces in Europe Treaty (CFE Treaty) 109–110
cruise missiles *see* nuclear cruise missiles
CTBT *see* Comprehensive Nuclear Test Ban Treaty (CTBT)

damage limitation 125–127, 131
Department of Defense 7, 12, 113, 119, 157, 179
deterrence 3–5, 8–9, 13–15, 29–30, 41–42, 194–195; China's views on 135–137; and Chinese nuclear strategy 70, 82n45; and Cold War theories 125–127; and excess 147–149; and international security 175–177, 182–184; limited 61–62, 64–67, **67**, 69, 72–75, 77, 134; minimum 61–69, **67**, 71–78, 116; and NATO nuclear modernization 106–108, 113–117; and the nuclear nonproliferation regime 92–93; post-Cold War 133–137; rethinking Cold War nuclear deterrence 130–133
dialogue 83n62, 120–121
disarmament 1–4, 7–9, 86–93, 101–102, 195–196; by default 157–158; *see also* disarmament, U.S.-Russian bilateral
disarmament, U.S.-Russian bilateral 40–43; and arms control and quantitative changes to the arsenals 50–53; future prospects for 53–54; and non-strategic nuclear weapons, doctrine 47–50; and strategic aviation 45–47; and strategic fleet 44; and strategic rocket forces 43–44
doctrinal questions 112–114

Europe 12, 14–16, 19, 32, 34; changing security environment in 108–114; and International Relations theory 130, 138; and international security 149, 153, 173–176, 181, 184; and NATO nuclear modernization 106–107, 115, 117–119, 121; and the nuclear nonproliferation regime 97–98; and U.S.-Russian bilateral disarmament 47–48, 55n14; *see also* European Union (EU)
European Union (EU) 98, 111, 128–129
excess 146–147; excess NNSA infrastructure 154; unnecessary 147–154; unsafe 158–164; unsustainable 154–158

fleet, strategic *see* strategic fleet
force structure 62–64, 74–77, 155, 160
France 3, 94, 106, 117–118, 174, 177, 184

GBSD *see* Ground-Based Strategic Deterrent (GBSD)
great power 5, 12–13, 93, 129–130, 136–137, 182–185; conflict 193–194; peace 174–176
Ground-Based Strategic Deterrent (GBSD) 17, 179

heavy bombers 6, 28–29, 32, 35, 45, 148

ICBM *see* intercontinental ballistic missile (ICBM)
India 1–2, 65, 78–79, 81n24, 84n93–94, 192; and international security 174, 177, 184–185; and the nuclear nonproliferation regime 88, 90, 94
INF Treaty *see* Intermediate-Range Nuclear Forces Treaty (INF)
intercontinental ballistic missile (ICBM) 6, 17–18, 21, 27–28, 30–31, 35; and Chinese nuclear strategy 76, 78, 83n76; and international security 151–152, 161–164, 178–179; and U.S.-Russian bilateral disarmament 41–46
intermediate-range ballistic missile (IRBM) 65, 81n25, 84n94, 85n98
Intermediate-Range Nuclear Forces Treaty (INF) 14–15, 97–100, 107–108, 111–112, 118–119, 159–160; demise of 98–99; INF killer 32–33
international order 174–178, 182–183
International Relations theory: Cold War theories 125–127; macro-theories 127–130; nuclear modernization's anticipated effects on 124–125; and nuclear proliferation 137–140; post-Cold War nuclear deterrence 133–137;

rethinking Cold War nuclear deterrence 130–133; the way forward for 139–140
Iran 12, 90, 147, 159, 177, 184–186, 195
IRBM *see* intermediate-range ballistic missile (IRBM)
Israel 1, 88, 94, 174, 177, 184
IW-I program 17–18

Japan 3, 65, 79, 149, 175, 177, 183

Kristensen, Hans M. 27, 29, 32, 49–50

leadership 86–87, 90–93, 115–116
legacy systems 25–29, 155–156
Libya 139, 177
long-range standoff missile (LRSO) 15, 17, 46, 95, 115–116; and international security 151–152, 161, 179
low-yield weapons 3–4, 14–16, 18–20, 49–50, 148–151; dangers of 160–161; and NATO nuclear modernization 113–114, 119–120
LRSO *see* long-range standoff missile (LRSO)

MAD *see* Mutually Assured Destruction (MAD)
Minuteman III 17–18, 94, 115, 153, 162, 178–179
MIRV *see* multiple independently targetable re-entry vehicle (MIRV)
modernization 5; and the changing security environment in Europe 108–114; as a determent to international security 145–146, 164–166; and international order 174–178; and NATO 106–108, 114–121; and the NPT 87–93; and other nations' nuclear forces 183–184; as a promoter of international security 173–174, 185–186; Russia 25–35; U.S. 12–21, 93–96, 101–102, 178–180; third wave of 155–157; threat posed by 180–183; *see also* excess; International Relations theory; *under* United States (U.S.)
multiple independently targetable re-entry vehicle (MIRV) 6, 17, 135, 148; and Chinese nuclear strategy 76–78, 83n78, 83n80; and Russia's nuclear modernization 27–28, 30, 36n14, 36n18; and U.S.-Russian bilateral disarmament 43–44, 51
Mutually Assured Destruction (MAD) 126, 130–133

National Nuclear Security Administration (NNSA) 14, 21, 94–95; excess infrastructure 154; and international security 154–155, 157, 160, 164
NATO *see* North Atlantic Treaty Organization (NATO)
NC3 *see* nuclear command and control (NC3) systems
New START *see* New Strategic Arms Reduction Treaty (New START)
New Strategic Arms Reduction Treaty (New START) 12, 18, 43, 92, 137; and international security 145, 147–149, 155, 159–160, 166, 170n70; and NATO nuclear modernization 111–112, 121; the nonrenewal of 99–101; and Russia's nuclear modernization 25, 29, 33–35
NFU *see* No First Use (NFU)
NNSA *see* National Nuclear Security Administration (NNSA)
No First Use (NFU) 65, 67, **67**, 70–72, 75, 80n10, 83n61; and international security 176, 182
nonproliferation 194–196; and U.S. nuclear weapons 176–178
non-strategic nuclear forces 52, 110; NATO 117–118
North Atlantic Treaty Organization (NATO) 9, 15, 19, 26, 30, 32, 34; and the changing security environment in Europe 108–114; and International Relations theory 138; and international security 148, 150–151, 153, 162, 174–175, 181; and NATO nuclear modernization 106; and the nuclear nonproliferation regime 98–100; nuclear policy 114–115; nuclear posture 114–121; purpose of NATO's nuclear weapons 106–108; relations with Russia 110–111; strategic nuclear forces and modernization plans 115–117; and U.S.-Russian bilateral disarmament 48, 50
North Korea 1–2, 88, 195; and international security 147, 152, 154, 173–177, 185–186; threat posed by modernization of 180, 182–183
Northrup Grumman 18
NPR 87–90
NTP *see* Nuclear Nonproliferation Treaty (NPT)
nuclear arsenals 5–8, 42–43, 51–53, 115–117, 130–131, 135–137; and international security 145–149, 155–158, 173–174, 177–182; role of U.S. nuclear arsenal today 130–133

nuclear command and control (NC3) systems 13, 178–179
nuclear cruise missiles 161–163, 181
nuclear forces 1–4, 6–9, 12–14, 29–30, 192–194; and Chinese nuclear strategy 61–64, 66–68, **67**, 70–74, 77–79; and international security 145–146, 148–149, 160–161, 163–166, 181–185; and NATO nuclear modernization 106–110, 112–113, 115–117; and the nuclear nonproliferation regime 93–101; and U.S.-Russian bilateral disarmament 40–47, 51–52; *see also* non-strategic nuclear forces; strategic nuclear forces
Nuclear Nonproliferation Treaty (NPT) 87–90; impact of U.S. leadership on 90–93
nuclear policy 8, 47–49, 91–93, 107–109, 114–115
nuclear posture 48–51, 109–110, 114–115, 117–118, 121; *see also Nuclear Posture Review* (NPR)
Nuclear Posture Review (NPR) 6–7, 12–16, 25, 30; and air-delivered nuclear weapons 16–17; and ballistic missile submarines 18–19; and Chinese nuclear strategy 61, 78; and international security 146, 149, 176, 179; and land-based ballistic missiles 17–18; and NATO nuclear modernization 110, 112–113, 116, 119–120; and non-strategic nuclear weapons 19–20; and the nuclear nonproliferation regime 86, 93, 96–97, 101; and nuclear weapons production complex 20–21; and U.S.-Russian bilateral disarmament 42, 46–50, 52
nuclear primacy 130–133, 140
nuclear proliferation 128–129, 176–177, 180; extended theoretical debates 137–139; *see also* nuclear proliferation regime
nuclear proliferation regime 86–87, 101–102; and the nonrenewal of New START 99–101; NPT and NPT regime 87–90; and the Trump administration 96–99; U.S. leadership and impact on the NPT regime 90–93; and U.S. nuclear modernization 93–96
nuclear strategy, Chinese 61, 77–79; and the academic debate 62–66; and conditions of use 70–72; and evidence of transition 66–77, **67**; and force structure 74–77; and objectives 67–69; and targeting 72–74
nuclear war 61–64, 68–69, 72–73, 96, 160; accidental 161–163; and Cold War theories 125–127
nuclear weapons 185–186; and great power peace 174–176; and the imperative of U.S. nuclear modernization 178–180; and nonproliferation 176–178; purpose of NATO's nuclear weapons 106–108; and the role of other nations' nuclear forces 183–185; the special role of U.S. nuclear weapons 173–174, 185–186; and the threat posed by China 180, 182; and the threat posed by North Korea 180, 182–183; and the threat posed by Russia 180–181; and U.S.-led, rules-based international order 174–178; *see also specific weapons*
nuclear weapons, air-delivered 7, 16–17
nuclear weapons, non-strategic 7–8, 14–16, 19–20, 138; doctrine 47–50
nuclear weapons production complex 7, 20–21, 95, 97

Obama, Barack *see* Obama era
Obama era 3–5, 7, 12–13, 40, 51, 138; and international security 146–147, 154, 157, 159–160, 165, 176, 179–180; and NATO nuclear modernization 110–111, 113, 119; and the nuclear nonproliferation regime 86, 91–93, 96, 99–100; recapitalization program 151–153
opportunity costs 148, 164
order *see* international order

Pakistan 1–2, 88, 90, 94, 192; and international security 174, 177, 184–185
peace 86–89, 173–176; great power peace 174–176
Pentagon 119–120, 155–157
People's Liberation Army (PLA) 68–69, 82n39, 83n80
PLA *see* People's Liberation Army (PLA)
Poland 111, 177
policy *see* nuclear policy
posture *see* nuclear posture
power *see* great power
proliferation *see* nuclear proliferation regime

realism 127–130, 165–166
recapitalization programs 146–148, 151–153, 155, 158, 165

redundancy 131, 151–153
Reif, Kingston 98
rocket forces, strategic *see* Strategic Rocket Forces
Russia 1–9, 12, 14–16, 18, 191–195; and Chinese nuclear strategy 61, 63, 66, 79, 84n88, 85n98; and International Relations theory 127; and international security 145–153, 155–156, 158–164, 173–175, 177, 179–185; legacy systems 26–29; modernization 25–35; and NATO nuclear modernization 106, 108–109, 111–115, 117–121; new developments in 29–33; and the nuclear nonproliferation regime 90, 94–101; relations with NATO 110–111; threat posed by modernization of 180–181; and U.S.-Russian bilateral disarmament 40–54, 55n14, 55n17; why developments in Russia matter 34–35; *see also* Russian Federation
Russian Federation 25, 40–41, 55n7, 99, 109, 112

security, international 7–10; and countries other than the U.S. 180–185; and excess 146–164; and the imperative of U.S. nuclear modernization 178–180; and international order 174–178; modernization as a determent to 145–146, 164–166; modernization as a promoter of 173–186; and unnecessary excess 147–154; and unsafe excess 158–164; and unsustainable excess 154–158
security environment 8–9, 121, 140, 147; change in Europe 108–114
show of force 134–135
SIPRI *see* Stockholm International Peace Research Institute (SIPRI)
SLBM *see* submarine-launched ballistic missile (SLBM)
SLCM *see* submarine-launched cruise missile (SLCM)
South Korea 3, 65, 149, 175–177, 183
SSBN *see* U.S. ballistic missile submarines (SSBNs)
stability, strategic 9–10, 158–159, 191–193
START *see* Strategic Arms Reduction Treaty (START/START I/START II)
Stockholm International Peace Research Institute (SIPRI) 4, 50
strategic ambiguity 133, 135–137
Strategic Arms Reduction Treaty (START/START I/START II) 26, 35, 108, 137; *see also* New Strategic Arms Reduction Treaty (New START)
strategic aviation 45–47
strategic fleet 44–45
strategic nuclear forces 41–45, 106–110; NATO 115–117
Strategic Rocket Forces 27, 30, 36n10, 42–44
strategy *see* nuclear strategy, Chinese
submarine-launched ballistic missile (SLBM) 16, 18–21, 27–28, 31, 35, 135; and Chinese nuclear strategy 78, 84n93; and international security 148–150, 152, 154, 160–161, 166, 178–179; and NATO nuclear modernization 110, 113, 115–117, 119; and the nuclear nonproliferation regime 86, 94, 96; and U.S.-Russian bilateral disarmament 41, 44, 46–47, 49–50, 56n34
submarine-launched cruise missile (SLCM) 14–16, 20–21, 47, 49–50, 52, 119–120; and international security 149, 151, 161–162
submarines *see* ballistic missile submarines

targeting 64–66, 72–75, 130–131
theory *see* International Relations theory
transition 7–9, 66–67, **67**; and conditions of use 70–72; and force structure 74–77; and objectives 67–69; and targeting 72–74
Trident missile 18–20, 115–116, 119–120
Trump, Donald *see* Trump administration
Trump administration 4, 6, 9, 12–16, 18, 20; and International Relations theory 138; and international security 145–147, 149–151, 153–159, 161, 165–166, 179; and NATO nuclear modernization 108, 111–113, 119; and the nuclear nonproliferation regime 86, 90, 93, 95–101; and U.S.-Russian bilateral disarmament 47, 51–52, 54; *see also Nuclear Posture Review*

uncertainty 135–137
United Kingdom 3, 94, 106, 116–117, 174, 177, 184
United States (U.S.) 1–6, 8–10, 191–195, 197; attitudes towards the bomb 133–137; and Chinese nuclear strategy 61–66, 72, 76–79, 81n25,

82n55, 83n62, 84n82; and excess 146–147; and great power peace 174–176; and international order 174–178; and International Relations theory 124–127, 129, 138–140; leadership of 90–93; modernization 12–21, 86–87, 93–96, 101–102, 178–180; and modernization as a determent to international security 145–146, 164–166; and NATO nuclear modernization 106–113, 115–121; and nonproliferation 176–178; and the nonrenewal of New START 99–101; and the NPT and NPT regime 87–90; and the role of other nations 180–185; and Russia's nuclear modernization 25–26, 29–35; the special role of U.S. nuclear weapons 130–133, 173–174, 185–186; and unnecessary excess 147–154; and unsafe excess 158–164; and unsustainable excess 154–158; and the Trump administration 96–99; and U.S.-Russian bilateral disarmament 40–43, 47–54, 59n96; *see also* Obama era; Trump administration

U.S. ballistic missile submarines (SSBNs) 7, 18–19, 31, 44–46, 69, 135; and international security 163, 178–179

war *see* nuclear war

warheads 4–6, 14–15, 17–18, 20–21, 25–28, 30–31; and Chinese nuclear strategy 75–77; and International Relations theory 125–126, 131–132, 135–136; and international security 145–150, 160–162, 164–166, 178–179; and NATO nuclear modernization 116–117, 119–120; and the nuclear nonproliferation regime 94–101; and U.S.-Russian bilateral disarmament 41–44, 47–48, 50–53

weapons, other 5, 118–120, 193–194

willingness to fight 134–135

World War II 98, 127–128, 156, 174–175, 184

W78 17–18, 21, 154, 164

W87-1 18, 154

Yars 30, 43, 83n76